INVESTIGATIONS IN GENERAL BIOLOGY

INTRODUCTION

The rapid accumulation of biological knowledge in the past two decades has presented a serious challenge to the organization and content of laboratory work in introductory biology. Many of the advances require sophisticated equipment in order to demonstrate the particular phenomenon or the many new techniques. These problems are compounded by the rapid growth in underclass enrollments which makes necessary the duplication or multiple purchase of expensive laboratory ware. The "knowledge explosion" presents another problem: Which facts should be illustrated? Does the selection of laboratory work reflect the kind of biology being investigated? Should laboratory work emphasize molecular biology? Should surveys of the biotic world be included? These problems center around a view of laboratory work as essentially illustration, demonstration, or technique.

Another view considers laboratory work primarily as process. That is, the laboratory work should be concerned primarily with how biology is investigated. The investigative approach should appeal to the natural curiosity of human behavior and contribute to the development of critical thinking on the part of the student. Investigation is the theme of this manual.

Critical thinking does not occur in a vacuum. Therefore, the content of the exercises in this manual was chosen on the basis of the following points:

1. The work should not only be part of an investigative process, but also should assist the student in understanding biological phenomena.
2. Whenever possible, the material should be chosen from the world of the student's experience. For this reason, there is a strong organismal bias to the selection of material for study.

Because process is emphasized rather than facts, introductory material deliberately has been kept minimal. Suggested readings are listed at the end of many of the exercises. These readings are intended to help the student explore on his own. For most exercises, reading references can be assigned from any of the modern textbooks. For this reason and because the emphasis is investigative, this manual can be used with any text.

Some of the exercises do have lengthy introductions and detailed directions. These materials are included when our experience has shown that the student needs a minimal foundation in order to make the investigation. For example, before a student can describe an animal he has never before seen, he needs some knowledge of the principles of animal organization. Therefore, a lengthy introduction of animal organization is provided and directions for dissection of an invertebrate and a vertebrate follow to provide the basis for the student's subsequent explorations. If the student already has the foundation, he may immediately embark on his unknown. One major advantage of the investigative approach is that students may proceed at their own rate. The statement, "I've done this before," no longer is applicable because the student may be directed into self study in an area where he has no previous specific knowledge and where he may explore as far as his initiative dictates.

Some of the exercises are not primarily investigative. These exercises are included because I believe that some topics may be taught better in the laboratory than in lecture; for example, mitosis and the classification of organisms (including both the study of kinds of organisms and of the principles of classification). Even these topics may be related to investigations and to independent work on the part of the student.

Many of the experiments are found in one form or another in other manuals. However, in most manuals the exercises are presented as a "show and tell" exercise. We hope that in this manual an investigative approach receives the main emphasis. The exercises in this manual appeared first in a preliminary edition in 1966. All of the exercises have been rewritten; many of them have had extensive modification. Most of the major changes were designed to further emphasize investigation. Many minor changes were made where our additional experience showed that clarification, correction, or amplification was needed.

Several of my colleagues have assisted with the preparation of the laboratory material. Dr. Delbert Shankel, Department of Microbiology, provided the *E. coli* experiment of Exercise 16. Dr. James Koevenig, Department of Botany, provided many helpful suggestions and criticisms. Mary Makepeace also provided many helpful suggestions. I wish to thank Arthur L. Buikema, Jr. and Norman J. Willems for the photomicrographs and Suzanne Batra for the drawings. I also wish to thank the laboratory teaching assistants who helped test these exercises through six years of trial and to thank those many students for their patient indulgence of our mistakes. I know that the coming generations of students will continue to disclose our shortcomings.

A final word to the student. The laboratory period is essentially a work period. Your work will proceed more efficiently if you have read the exercise and given it some thought before coming to the laboratory. Because the introductions are brief, you may wish to read appropriate material in your text in order to increase your understanding of the laboratory work and to make it more relevant to your course of study. You may wish to do some preliminary planning so that you can proceed more rapidly and efficiently once you are in the laboratory. Blank pages and blank spaces have been included so that you may record your ideas and experimental plans. You are urged to sketch plans for experiments even though you do not have time to conduct the work. Such activities will sharpen your critical thinking. The many suggestions for investigation permit you to explore and learn to whatever extent you wish to devote your time and effort. In the final analysis, the amount of profit will depend, in large part, on the amount of investment.

KENNETH B. ARMITAGE

Lawrence, Kansas

EXERCISE

1 OBSERVATION AND DESCRIPTION

The simplest form of biological investigation is the qualitative description of a series of events of which you are the observer. This form of investigation is familiar because you use it in the conduct of your everyday life. You note if a traffic light is red or green; you may make a mental record of the hair color of a nearby student. Many other such qualitative observations will readily come to mind.

Indeed, much of our active life is spent making observations and ordering them into a meaningful system. Sometimes the ordering may be almost subconscious. At other times the ordering of information may require arduous effort. In finding your way to the laboratory, you had to order a series of observations involving left or right turns combined with walking upstairs or downstairs. The entire series of independent events becomes ordered in your brain so that after a few attempts (or maybe one) you can find your way to the laboratory automatically. The ability to obtain information from the environment and to order the information into significant patterns is a characteristic of living organisms. But organisms differ in the kinds and complexity of information they can obtain and order.

Man probably has the highest capacities for ordering information. This ability must have conveyed great survival value to those individuals in whom observation and integration were well developed. Undoubtedly for early man such ability was a matter of life or death. Even today, critical observation is vital to survival in those human societies living in harsh environments. The Bushmen of the Kalahari Desert are a hunting, food-collecting culture. They have an immense knowledge of the plant and animal life of their environment. The skill with which they can find food in their harsh homeland amazes visitors. Most members of our western culture would soon die in the same area in which the Bushman lives. Does this difference in ability to survive imply that the Bushman is more intelligent? Not necessarily. It does tell us that the Bushman has observed and ordered his environment in such a way that he can survive in it. Likewise, we observe and order our environment so that we may survive as long as we can. The essential difference between the

Bushman and us is the kind of information that is ordered and the tools that are used to gain information. Let us emphasize again that the ordering of information is basic to human behavior and is basic to science.

In this sense, science is an extension of basic human activity. There are various definitions of science and most elementary textbooks have an introduction to the nature of science and the scientific method. You may be aware that we are formulating a definition of science based on natural human behavior. Let us, then, define science as the systematic ordering of information. In our culture, science includes not only the ordering of information necessary for survival and comfort (applied science), but also those researches that may be characterized as "pure" science. Usually we think of "pure" science as the ordering of information for its own sake.

Our objective is to examine the means by which the systematic ordering of information may occur. We began this discussion with the statement that the description of events is a simple ordering of information. The degree of orderliness depends upon the accuracy of the observation. As we shall see later, orderliness also depends on the interpretation of the events recorded. But for the present, we wish to stress the importance of accurate observation.

1. BEHAVIOR OF THE PARADISE FISH

For this exercise, we have placed two male paradise fish, *Macropodus opercularis,* in adjoining compartments of an aquariam. The compartments are separated by means of an opaque partition. You should begin your description of the behavior of these fish by carefully recording the conditions under which the fish are maintained.

Your objective is to describe the behavior of the fish when the partition is removed. We have not provided you with any description of the typical pattern of behavior, nor any clues as to what to expect. Nor have we provided you with any directions for recording your observations except the following.

Take notes on the behavioral activities of the fish. Record everything that you think may be of any significance.

After you have completed your observations, order your information into a qualitative description of the behavior. You may wish to give names to any of the behavioral events you described. You may also wish to interpret or give meaning to the behavior. What kinds of questions would you ask and why?

DISCUSSION QUESTIONS

The name, *Macropodus opercularis* L., is the scientific name of the paradise fish. Why do you suppose scientific names are used?

Did you name or categorize any of the behavioral events you described?

What would be the purpose of naming particular aspects of the fish's behavior?

2. BEHAVIOR OF FLY LARVAE

For this study, a larva of the house fly, *Musca domestica*, or of another kind of fly is removed from a darkened container and placed on slightly dampened filter paper in the bottom of a petri dish or other container which is lighted from above by diffuse light.

Describe the locomotor activity of the fly larva. Can you discern any pattern to the movements?

1. *Coarse adjustment.* This knob is the larger of the two wheels and is the upper one.

2. *Fine adjustment.* This knob is smaller and is the lower of the two.

The coarse adjustment is used to focus the lens system with the low power or the scanning objective. Once the object is focused as sharply as possible, the fine adjustment may be used to refine the focus. When the high power objective is used, only the fine adjustment should be used to focus.

C. Lighting

1. *Mirror.* The mirror is located below the object and directs light rays from the source of light to the object. Normally the mirror is adjusted so that the maximal amount of light is directed on the object.

2. *Disc diaphragm.* The disc diaphragm contains a series of openings of different size. These apertures control the amount of light reaching the object. In general, one should use an aperture of such size that the field is evenly illuminated.

NOTE: Your microscope may be equipped with an *iris diaphragm* which may be opened or closed slowly to any size aperture. Additionally, there may be a *condenser* which may be raised or lowered to bring the light from the mirror to a focus on the specimen.

D. Support

1. *Base.* The horseshoe-shaped foot with its pillar supports the microscope.

2. *Stage.* The stage is the working platform on which the object is placed for observation. Usually the stage is provided with one or more clips for holding microscope slides steady.

3. *Arm.* The arm holds the lens system in the proper position above the stage. Whenever the microscope is carried, one hand should grasp the arm while the other hand supports the base.

2. USE OF THE MICROSCOPE

The main problem in the use of the microscope is proper focusing. Proper focusing involves the correct alignment of the lens system and a sufficient amount of light.

Place the microscope in front of you with the open end of the horseshoe pointing toward the source of light. Turn the nosepiece so that the low power objective is aligned with the ocular. Using the coarse adjustment, turn the objective downward as far as it will go. Then adjust the mirror so that light is reflected through the lens system to your eye. You should see a bright circular field with a black line extending into it from one edge. The black line is a pointer which is fastened in the ocular. The ocular may be rotated so that the pointer may point to a specific area in the field. Whenever you wish to ask a question about some structure in the field, always identify the structure with the pointer so that you and the instructor are looking at the same thing.

Place the slide marked "e" on the stage. Center the slide so that the "e" is under the objective. The light should pass through the "e". Slowly turn the objective upward until the "e" comes into focus. Note that the "e" is inverted. Move the slide to the right. Which way does the letter move?

Move the slide toward you. Which way does the letter move?

Because of the inversion of the image, movement will be in a direction opposite to what you expect. It is important to become accustomed to this reversal so that you can readily follow live, moving organisms.

The ability to bring an object into sharp focus is not a function of magnification, for there is no theoretical upward limit to the magnification of a microscope. Focus is limited by the resolving power, the ability of the microscope to render visible the detail of the object. Once an object becomes fuzzy because of the limitations of the resolving power, further magnification only makes the image larger and less distinct. The resolving power depends on the design of the objective. Essentially, the objective which picks up a larger cone of light has a better resolving power and gives more detail in the image than an objective which picks up a smaller cone of light. There is, of course, a limit to the resolving power of the light microscope and better resolving power is obtained by means of the electron microscope. In focusing, remember these rules:

1. Find the object first under low power. *Never* search for an object under *high power*.
2. *Never* focus *downward* with the coarse adjustment. This rule is especially important when using the *high power objective* because the objective is long enough that it can be run into the slide, and the lens or slide, or both, can be scratched or broken.
3. When focusing under high power, use only the fine adjustment.

Switch the nosepiece so that the *low power objective* is aligned with the ocular. Now move the "e" to one side of the field. Focus sharply. Then switch back to the *high power objective*. Depending on where you placed the "e" you may see only a small part of it or none at all. Therefore, if you are examining an area or object under low power and wish to change to high power, the object or area should be centered in the visual field before increasing magnification. As magnification increases, the size of the field decreases. It is well to have some idea of the size of the object you are studying. A good approximation can be made if you know the diameter of the field and how much of the field the object occupies.

Examine the metric scale on your ruler under the *scanning objective*. When the scanning objective is in focus, the body tube will be raised higher than when low or high power objectives are in focus. Carefully measure the diameter of the field. Record the measurement in Table I. Do the same for the *low power objective*. The high power objective magnifies too much for direct measurement. Since the high power objective magnifies about 4.3 times more than the low power objective, the diameter of the field is reduced by about 4.3. Calculate and record the diameter of the field for the high power objective.

When you use the microscope, it is preferable *not* to *bend* the arm. If the arm is bent, wet mounts will tend to run to the downslope side. Adjust your stool so that you sit comfortably and look through the microscope without

TABLE I

Name of objective	Magnification	Color code	Diameter of the field
High power	43×	Yellow bands	
Low power	10×	Green bands	
Low-low or scanning	2×	No bands	

tilting it. When you put the microscope away, be certain that the *low power objective* is aligned. *Never* put the microscope away with the high power objective aligned. The lens should be cleaned each time before you start using the microscope. Use *lens paper only*.

Again center the "e" in the field and focus as sharply as possible. Turn the nosepiece and align the *high power objective* with the ocular. The "e" will be much larger and almost in focus. The letter is in or near focus because the objectives are parfocal. Any final focusing should be done with the *fine adjustment*.

The stage clips help stabilize the focus. Ordinarily, once the object is located, the slide is fastened by means of the stage clips and the microscope focused.

Practice inserting slides and focusing the microscope until you can do so with facility. Your instructor may supply living organisms for your observation.

SUGGESTED READINGS

Lenhoff, Edward S. (1966). "Tools of Biology," 120 pp. MacMillan, New York. Paperback.

EXERCISE

3 QUANTITATIVE DESCRIPTION AS ANALYSIS

In Exercise 1 we examined biological phenomena as a series of observed events. Additionally, we modified the conditions of observation, but we had no particular plan or reason for choosing the modifications that were used. In a sense, we were saying "I wonder what would happen if . . .". Many interesting biological facts have been discovered by means of this expression of curiosity. But usually when a modification or experiment is tried, the investigator has some reason for doing so. It may be only a vague idea formed in the experimenter's brain or it may be a carefully constructed logical system. Essentially, the investigator has some observations from which he formulates an idea or hypothesis. Then the hypothesis is tested. The test results may confirm, deny, or cause a revision of the original hypothesis. We shall examine diffusion as a simple example of this system.

Diffusion is defined as the movement of molecules away from the area of their highest concentration. The potential validity of the definition may be assumed from common experience. We are capable of smelling many substances and it seems evident that something from the substance is reaching our smell receptors. Obviously, the substance at its source is more concentrated than when it reaches our noses. You may recall that an odor may get stronger as you get closer to its source. Does this observation mean that the substance becomes more concentrated closer to its origin?

Let us now look at a series of observations which will enable us to derive some characteristics of diffusion.

1. DEMONSTRATION OF DIFFUSION

Fasten a long glass or plastic tube horizontally to a ring stand. Moisten two pieces of cotton, one with hydrochloric acid (HCl) and the other with ammonium hydroxide (NH_4OH). Plug one end of the tube with one piece of the moistened cotton and plug the other end of the tube with the other piece. Carefully watch for the appearance of white smoke (NH_4Cl). Where in the tube does the white smoke first appear?

We may ask several questions about this observation.

Did diffusion occur?

The appearance of the white smoke may be accepted as evidence that molecules have diffused from the moistened cotton to react to form the smoke. Probably when you moistened the cotton you smelled the ammonia or the hydrochloric acid or both, and this experience provides further support that diffusion did occur. But you may wish to verify that HCl and NH_4OH formed a white precipitate when they came into contact. Can you suggest and perform a simple test to verify this point?

Why did the white cloud form closer to the end of the tube containing the HCl?

The obvious answer to this question is that the ammonia must have diffused faster. But why should ammonia diffuse faster than hydrochloric acid?

What properties of ammonia and of hydrochloric acid are related to rate of diffusion?

At this point we might form a hypothesis if we had some additional information. But what information do we need? We might guess that the abilities of these compounds to react with other kinds of substances probably would have no bearing on diffusion. And we might search through other properties of these substances looking for a clue. We might eventually note that the molecular weight of ammonia is 17 and of hydrochloric acid, 36. From this information we could formulate the following hypothesis: The rate of diffusion is dependent on the size or molecular weight of the diffusible molecules.

2. TESTING OF THE HYPOTHESIS CONCERNING THE RATE OF DIFFUSION

If the hypothesis that the rate of diffusion is related to molecular weight is correct, then the measured rates of a series of molecules should be proportional to their weights.

The testing of this hypothesis seems obvious. Simply measure the rates of diffusion of a group of molecules. But note that we have added a consideration which was not present in our demonstration of diffusion. We need to quantify the rate at which molecules move. Since we cannot see molecules as such, how can we measure their movement? One possibility is to measure the molecules indirectly or to use molecules that can be seen en masse (e.g., the NH_4Cl). Should these tests be conducted in air, water, or some other medium? In other words, we need to find a system in which our hypothesis can be tested. This consideration is critical to the design of biological investigations. The ordering of information is always related to the conditions under which the information is obtained.

The use of air or water as the medium in which diffusion occurs has obvious difficulties. How are air movements controlled? What sort of index of rate of movement can be devised? The problems in water are much the same. If the substance dissolves and cannot be seen, how can the rate of diffusion be measured? Also, the substances must diffuse in the medium that is used. In summary, substances that can be seen as they diffuse and a medium which will permit diffusion, but not affect the rate of diffusion, are necessary to test the hypothesis.

The first problem in design, that of detecting the diffusing substances, may be solved by using colored compounds. If, additionally, these compounds are soluble in water, then water can be the medium in which diffusion occurs.

However, diffusion in water is difficult to measure, as can easily be determined by dropping a crystal of potassium permanganate into water. Diffusion is rapid and irregular and slight movements of the container of water disturb the pattern of diffusion. What is needed is some way to steady the water. This problem may be solved by incorporating the water into a gel. A gel consists of a framework of solid particles which trap water in their interstices. (Gelatin is an excellent common example of a gel.) Thus, diffusion may occur in water, but the water itself will not move about and influence the rate of diffusion.

With the above points in mind, the following experiment may be performed. The gel which is used is an agar-water gel in a petri dish. Agar is a protein obtained from certain seaweeds which readily forms a gel when a small amount is added to hot water and the mixture cooled.

WARNING: Potassium dichromate is a harmful dust which may cause skin rash or external ulcers. Do not get this substance on the skin or in the eyes. If some does come in contact with the skin or eyes, wash the area thoroughly with water. There is no danger if normal laboratory care and decorum are followed. Handle all the crystals with a pair of forceps.

Select one crystal each of potassium permanganate, methylene blue, and potassium dichromate of about the same size. Place each crystal carefully on the surface of the agar, noting which is which so that the crystals are spread equally over the agar and not too close to the edge of the dish. Measure the diameter of the colored area for each crystal immediately after adding the crystals to the agar and record the measurement in millimeters (mm) for 0 minutes (min). Using a regular time interval (how can you determine what interval is best), measure the diameters of the colored zone around each crystal and record both the time (in minutes since 0 time) and diameters in the appropriate columns in Table I.

One of the conditions which must be met in performing this experiment is that the compounds used must be soluble in water. The importance of this condition may be tested by adding a crystal of manganese dioxide to the agar. Manganese dioxide has a molecular weight of 87, but is insoluble in water. If its insolubility were unknown, a possible conclusion might be that diffusion

TABLE I

Time	Potassium permanganate (mol wt 158)	Potassium dichromate (mol wt 294)	Methlyene blue (mol wt 374)
0 min			

How would you determine how many measurements to make?

is more rapid over the middle range of molecular weights and decreases as molecules get larger or smaller.

The next important factor that we must consider is organization of the data gathered in the experiment. The data should be ordered in such a way as to clearly illustrate the correctness or falsity of the hypothesis. The hypothesis is "the rate of diffusion is dependent on the size or molecular weight of the diffusible molecules." The most direct means of evaluating the data is to inspect the table and see if, indeed, the potassium permanganate diffused the greatest distance and methylene blue, the least. However, such inspection does not easily enable us to find any precise relationship between diffusion and molecular weight. A clearer relationship frequently may emerge when data are graphed. For this experiment, the most direct relationship may be shown by plotting the final diameter of the colored spot against molecular weight. Fill in the following graph using the values obtained in your experiment:

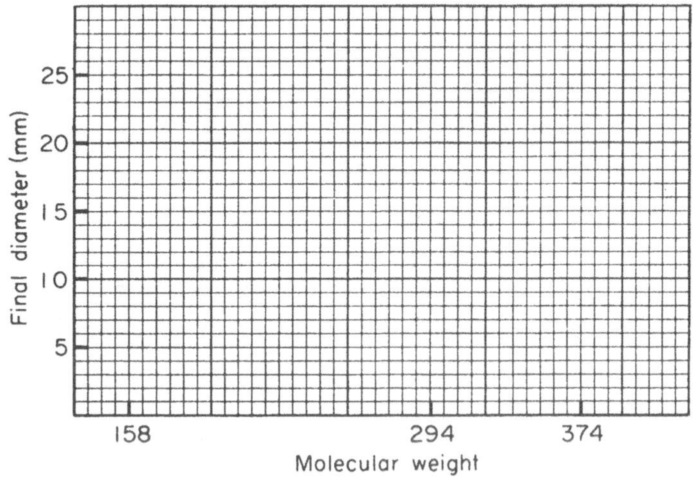

If the rate of diffusion under those experimental conditions is closely related to molecular weight, the graph should approximate a straight line. Therefore, we can state that out hypothesis is acceptable.

Does this experiment prove that the rate of diffusion is always closely related to molecular weight? Does this experiment increase the probability that the hypothesis is valid?

Frequently data may show relationships that were not anticipated when the experiment was designed. The manner in which the data are analyzed is crucial to the amount of information that may be gained. Let us examine our data from this experiment as an illustration of this point.

Our plot on the above graph has a low information content because only one possible relationship can be examined. In our data collecting, we measured the diameter of the colored circle at regular intervals. Because we have a time division, we may ask the question, "Is there any relationship between time and the rate of diffusion?" To test this possibility, we can plot distance (the diameter measurements of each diffusion area) against time. In effect, we are formulating another hypothesis: the rate of diffusion is independent of time.

Plot distance (on the vertical axis) against time (on the horizontal axis) on graph paper for each of the three substances.

Do you accept or reject the hypothesis that the rate of diffusion is independent of time?

Does this type of plot enable you to decide if the rate of diffusion is related to molecular weight?

Is the relationship as evident as in the previous graph?

Can one always expect to obtain all the information from a set of data by only one treatment of the data?

At this point you should be aware that the ordering of information requires not only looking for the expected but being alert for the unexpected. Examine once again your petri dish of agar with its diffusion patterns. Are there any other characteristics of diffusion that you can observe? List those characteristics below.

Several features of biological investigation have been presented and discussed in this section. A summary of these features may now be appropriate to focus your attention on the investigative process.

1. A specific observation may suggest a more universal principle or relationship.
2. The relationship or principle may be formulated as a hypothesis.
3. The testing of the hypothesis requires appropriate materials and methods.
4. The data obtained from the test of the hypothesis must be analyzed in a manner which will give maximal information concerning the question asked.
5. The data may contain information demonstrating relationships in addition to or in place of the principle being tested. Thus, the original hypothesis may be accepted or rejected and one or more new hypotheses may be formulated.
6. The new hypothesis may be tested by data already available, or may require further testing.

The summary should be qualified in one important respect. Do not assume that every investigation will result in data that will suggest new relationships. Probably the great majority of experiments are designed to test the validity of a given hypothesis and the results do nothing more than increase or decrease the probability that the hypothesis is correct.

3. PRESENTATION OF RESULTS

Whatever the nature of a scientific investigation, the results must be presented accurately, clearly, and concisely to other persons who are interested in them. These may be a small group of specialists or may be all persons who can read (to cite the extremes). Essentially, writing a scientific report does not differ from any other writing in which the purpose is to convey information or ideas. In form, scientific writing usually has the following organization:

Introduction. This section includes the rationale for the investigation. There may be a brief history of published reports that recounts the events that led to the work that is being reported. There is a statement of purpose—why the author performed this particular work. In a sense, the purpose may contain the hypothesis or hypotheses being tested even though there may be no formal statement of a hypothesis.

Materials and methods. This section describes the conditions under which the work was performed. These conditions should be described in such a way that the reader can duplicate the work in his own laboratory.

Results. The data obtained in the research are presented in an organized manner. Usually data are summarized in tabular or graphical form or may be described in mathematical notation.

Discussion and conclusions. Essentially, this section states reasons for accepting or refuting the hypothesis. In practice, the interpretation and implications of the data are discussed. Other studies may be cited and their relevance to the present report pointed out. Suggestions for future investigations may also be included. This section does not involve a restatement of the data, but references to the data occur as the author pieces together a logical interpretation of the results of the investigation.

Prepare a written report of the experiment testing the hypothesis that the rate of diffusion is related to the molecular weight of the diffusible substances. The report should follow the form outlined above and should be based primarily on the laboratory work. However, at the discretion of your instructor, the discussion might include reference to literature sources (such as your textbook). If reading materials are used in the report, the style manual of the American Institute of Biological Sciences (AIBS) should be used as a guide to the proper ways of citing literature sources and making a bibliography.

This paper is requested to give you experience in developing the form and content of these reports. In subsequent exercises, the report will inform the laboratory instructor of the nature of your laboratory work and of your abilities to synthesize and interpret data.

SUGGESTED SUPPLEMENTARY EXERCISE

You may read articles in several scientific journals. Preferably, this reading should be done in the library. You will become familiar with the library, learn where scientific journals are kept, and determine something about the divisions of biology into subdisciplines as expressed in journals.

You might analyze several papers. Are the papers descriptive or analytical? Are they qualitative or quantitative? Did the author(s) clearly state a hypothesis (even though the term hypothesis was not used)? Are the conclusions clearly stated? Can the reader determine what the author(s) did?

SUGGESTED READINGS

Gabriel, M. L. and Fogel, S. (1955). "Great Experiments in Biology," 317 pp. Prentice-Hall, Englewood Cliffs, New Jersey. Paperback of collected readings.

Hempel, Carl G. (1966). "Philosophy of Natural Science," 116 pp. Foundations of Philosophy Series. Prentice-Hall, Englewood Cliffs, New Jersey. Paperback. This book provides an extensive discussion of the structure and testing of hypotheses, the nature of scientific laws, theories, and concept formation.

Knobloch, I. W. (1967). "Readings in Biology Science," 491 pp. Appleton-Century-Crofts, New York. Paperback of collected readings.

Kormondy, E. J. (1966). "General Biology." Vol. I: "Molecules and Cells," 252 pp. Vol. II: "Organisms, Populations and Ecosystems," 241 pp. Brown, Dubuque, Iowa. Paperbacks of collected readings.

Laetsch, W. (1969). "The Biological Perspective," 574 pp. Little, Brown, Boston, Massachusetts. Paperback of introductory readings.

EXERCISE

4 USE OF MODELS

Many complex phenomena are more readily analyzed if they can be reduced to simpler terms. This simplification may involve the formulation of a model. The model may be mathematical, may be a series of diagrams on paper, or may be a working physical system.

A model may be defined in two ways. It may be anything used as a guide in testing a new design, such as a small-scale copy of a bridge or airplane. A model may also be a design or structure representative of some system. In biology, the latter definition of a model is the one most generally used. The model in this sense becomes a hypothesis for testing, so that the validity of the model may be established or rejected. Thus, a model provides a logical framework for analysis.

A model may stimulate new ways of thinking about a problem and suggest new experiments to test the validity of the hypothesis. A model may have a greater or lesser identity with the real situation. It may begin as an abstraction or analogy and eventually become identical to a precise description of the system. For example, the solar system was at one time thought of as a series of bodies in orbit about the earth. But as knowledge accumulated, that model no longer satisfied the known conditions and was replaced by one in which the sun became the center of the solar system. This model seems to provide good agreement with known facts and we accept the hypothesis of the heliocentric solar system. The validity of the model may be further attested to by our success in orbiting space vehicles around the moon and sun. Thus, the model comes to serve as a description of reality as we know it. A more recent example from biology may illustrate further. The structure of deoxyribonucleic acid (DNA) was the object of intensive research. Several models were proposed, but none withstood testing until the Watson–Crick model was formulated. Today the Watson–Crick model seems to be an accurate description of the structure of DNA. In summary, one use of a model is that it becomes the best description of the system that the investigator can devise.

A second use of models is less precise. The model represents a broad concept or general principle which provides a fruitful and interesting way to explore certain problems. The model is useful only as long as it stimulates research and analysis. The research is not directed toward establishing the validity of the model as such, but explores the implications suggested by the properties of the model. One such model is a feedback control system for maintaining steady state or homeostasis of the internal environment of organisms. This model in its simplest form consists of a receptor, information pathways, integrator, and effector:

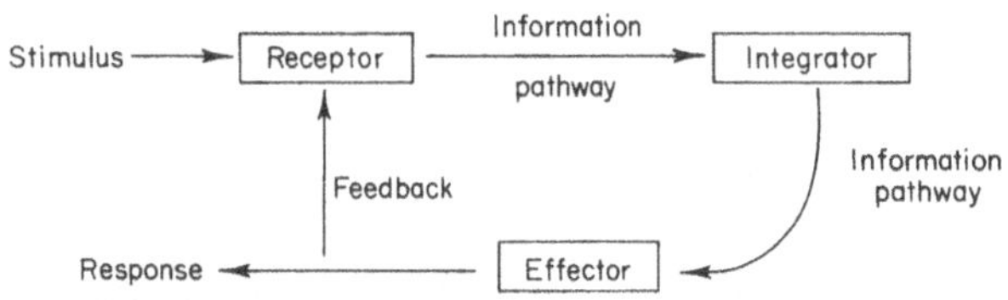

Briefly, an environmental change or stimulus is detected by a sensing element or receptor which sends information to a structure capable of integrating incoming information and selecting a response. The information which brings about the response is sent to an effector which carries out the response. The response provides information as feedback to the system so that the system can adjust its response pattern to maintain homeostasis. The fitting of physiological data to the model in a precise way is relatively difficult if for no other reason than the great complexity of the system. Of course, the model may be modified and made more complex so that it becomes a description of the physiological events and morphological components which regulate a system (e.g., blood pressure, breathing). But each system would require a somewhat different specialized model and would tend to obscure the general utility of the simple model. Thus the simple model is useful because it provides a way of thinking about biological systems and because it provides a general framework to which specialized data may be related.

1. USE OF A MODEL TO STUDY A CELLULAR PHENOMENON

Models have been especially useful in the study of cells because of the difficulty of investigating such small entities. Thus much of the description of cell function was based on the ways certain models worked. The model came to function as a hypothesis which could be tested. The analysis of the function of the cell membrane is particularly representative of the use of models.

In Exercise 3 we investigated some of the properties of diffusion. Diffusion is particularly important in biological systems because it is a basic transport mechanism. Cells must obtain materials from their environment and must eliminate other substances into their environment. The passage of materials must cross the cell membrane. Chemical analysis shows that the internal and external environments of cells differ. Thus the cell membrane was thought of as a structure capable of regulating the movement of substances into and out of the cell. By what means could this regulation occur? A simple hypothesis is that the cell membrane is differentially permeable (or semipermeable). That is, some materials may not pass through the membrane at all and other materials pass through at different rates. Because water passes in and out of cells so easily, our model should include a membrane freely permeable to water. The validity of the model of a differentially permeable membrane may be tested by first determining the properties of the model and comparing these with the properties of membranes of living cells.

If we are to describe some properties of a model of a cell membrane, we should choose a relatively simple procedure. Especially should we establish conditions under which we would expect some substances to diffuse. Such a system may be established if the concentrations of one or more substances differ on the two sides of the cell membrane.

A. Preparation of the Model

1. Obtain a piece of dialyzing membrane.
2. Close off one end by doubling back a half-inch portion and tying it securely.
3. Fill the sac with 1 molar (*M*) salt (sodium chloride) solution to which methylene blue (molecular weight 374) was added.

NOTE: A mole is a gram molecular weight of a substance. Thus, a molar solution of sodium chloride contains 58.4 grams (gm) per liter of water.

4. Tie off the other end (as in 2 above) to form a closed cell. Try to eliminate all the air. Do not fill the cell too full; it should be somewhat flaccid.
5. Weigh the cell. Weighing may be done directly on a balance or by measuring the displacement of water in a graduated cylinder. Assume that 1 milliliter (ml) of water is equal to 1 gm at room temperature.
6. Carefully rinse a clean container with distilled water, then fill with fresh distilled water.
7. Immerse the cell completely in distilled water. Remove, immediately, a few milliliters of distilled water from the container with the cell. Add 1 drop of silver nitrate solution to this sample. If sodium chloride is present, a white precipitate of silver chloride will form. If your test for sodium chloride is positive, what should be your next course of action?

8. At regular intervals weigh the cell and test the water around the cell for the presence of sodium chloride. Follow the procedures outlined in 5 and 7 above. Record your data in Table I.

TABLE I

Time (since immersion)	Weight of cell	Sodium chloride		Methylene blue	
		Absence	Presence	Absence	Presence
0 min					

9. Also record the time of any appearance of blue color in the solution bathing the cell.

Is the membrane of the model cell differentiably permeable? Cite evidence from your data to justify your answer.

Another question that might be asked of the model is whether water will diffuse out of the cell as well as in. This question can be answered by constructing a cell (as in 1,A above) and placing it in 2 *M* sodium chloride solution. Weigh the cell several times over a period of 30–40 min. What conclusions can you make about the movement of water as a result of this experiment?

Can you now predict the change in size that would occur if the cell model is placed in a solution of the same molarity? State a hypothesis and conduct an experiment to test the hypothesis.

Living cells contain macromolecules, essential constituents which are missing from the model. The model may be refined by conducting an experiment in which the cell constructed of dialyzing membrane is filled with colored gelatin (gelatin is a protein). What happens to such a cell? Record your data and conclusions in the following space.

Finally, describe the hypothetical properties of a cell membrane based on the results from the above experiments. In other words, you will use the data obtained from working with the artificial cell membrane to construct a model of the living cell membrane. This model should include some predictions about what might happen to cells when they are placed in solutions of different concentrations. Some of these predictions will be tested in the next exercise.

SUGGESTED ADDITIONAL ANALYSES OF THE CELL MODEL

One question which you might ask is whether the rate or amount of diffusion across a membrane depends on the concentration of the cellular fluids. This question will require highly accurate measurements of diffusion.

Was the method used in the experiment just performed sufficiently accurate? Can you suggest alternative ways of making more accurate measurements of diffusion across the membrane?

Can diffusion occur at the same rate in either direction across the membrane of the model? Does the same relationship occur in living cells?

Does it make any difference what kind of salt solution is used in the model?

What happens to the weight of the model when it is left in the distilled water for several days? What happens to the concentration of sodium chloride inside the cell relative to the concentration outside the cell? Does the same phenomenon occur in living cells?

SUGGESTED READING

Langley, L. L. (1965). "Homeostasis." Reinhold, New York.

EXERCISE

5 TESTING THE MODEL OF THE CELL MEMBRANE

The validity of the model of the function of the cell membrane may be tested first by examining cells under the microscope and recording their responses in solutions of different concentrations.

Before discussing an experiment to test the cell model, it may be appropriate to digress briefly and describe some of the characteristics of osmosis. *Osmosis* is defined as the passage of a solvent, such as water, across a differentially permeable membrane. When the space on one side of the membrane is limited in volume, the force of solvent flow in the direction of higher solute concentration may exert a pressure (*osmotic pressure*). Osmotic pressure may be measured, but ordinarily it is easier to examine it indirectly. Any cell is a solution of water and solutes. Any solution will have an "attraction" for water depending, in part, on the concentration of the dissolved substances. This attraction for water is *osmotic potential*. The osmotic potential usually may be measured indirectly by measuring the *osmotic concentration* of the solution. The osmotic concentration is the number of dissolved particles, such as molecules and ions, which contributes to the directional movement of water. Osmotic concentration usually is expressed as moles per liter.

Refer to your records on the response of the model. When placed in solutions of higher molarity (and hence higher osmotic concentration), the model lost weight. The opposite reaction occurred when the model was placed in a solution of lower osmotic concentration and there was no change in weight in a solution of the same osmotic concentration. This experiment indicated that water may move in either direction through a membrane, but that the net movement is zero if there is no difference in osmotic concentration. Does water move in either direction through a cell membrane? We may answer this question by performing the same kind of experiment with living cells.

A. Responses of Red Blood Cells to Solutions of Different Osmotic Concentrations

Ideally, the experiment should be conducted as if the osmotic concentration of red blood cells were unknown. However, in order to save time, we may note that red blood cells have an osmotic concentration equal to about 0.15 *M* sodium chloride. This information permits us to determine a starting point for comparing the responses of cells with the responses of the model. When we characterized the model, we used the initial weight of the membrane sac as the starting point for measuring any subsequent changes. The red blood cells are too small for weight to be used as a measure of change, but we can use a measure of turgidity. Just as the dialyzing membrane sac became turgid in a dilute environment, so should the red cell membrane stretch if our model is relevant to a cell. The cell membrane stretches because of the uptake of water and subsequent increase in cell volume. This change may be measured by determining the diameter of the red blood cells. The conspicuous change occurs in the general shape of the cells. Normally, the cells are biconcave discs, a condition seen easily when the cells are viewed from the edge. In solutions in which the osmotic concentration is lower than that of the cells, the cells tend to become biconvex.

If large differences in the molar concentrations of the salt solution are used, the differences in the diameter of the red blood cells are easily evident to the eye. However, more accurate measurements are more desirable and can be made with an ocular micrometer. The ocular micrometer need not be calibrated for purposes of comparing cell sizes. Express the cell diameter as the number of units (vertical lines) on the ocular micrometer (a cell may have a diameter of 20 units, 19 units, etc.).

You have been asked to accept as correct the statement that red blood cells have an osmotic concentration equivalent to 0.15 *M* sodium chloride. As you study the responses of the red cells in the subsequent tests, you might think of how the data you collect might verify the validity of this standard osmotic concentration.

We may now begin to test the validity of the model. Because we measured responses of the model in three molar solutions, we can make a quick comparison by measuring red blood cells in three molar solutions. One saline solution should be higher than, and one should be lower than, 0.15 *M*.

1. Obtain three clean microscope slides.
2. Label one slide 0.07 *M*, one slide 0.15 *M*, and one slide 0.30 *M*.

3. Wash the end of one finger with a cotton ball dipped in alcohol. After the alcohol dries, prick the finger with a sterile lancet.
4. Place on each microscope slide one small drop of blood.
5. Immediately add saline solution to each drop of blood. Be certain that the concentration of saline added agrees with the label on the slide.
6. Cover each drop of blood and saline with a cover slip. Place one edge of a cover slip in contact with and angled over the drop of blood and gently lower the cover slip into the solution (allow the cover slip to drop the final distance).
7. Examine each slide under high power on the microscope.
8. Describe the reaction of each set of cells in each solution.

NOTE: The solution under the cover slip will slowly become more concentrated as water evaporates. Add more solution to the edge of the cover slip as necessary to prevent a change in concentration.

The red blood cell is a specialized cell of some animals. We might question whether the results obtained using this cell are applicable to a general test of the validity of the membrane model. Therefore, the model should be tested on another kind of cell.

B. Responses of Plant Cells to Solutions of Different Osmotic Concentrations

Elodea is a flowering plant that grows with most of its parts under water. This widespread plant is commonly used in aquaria and is easily cultivated in the laboratory. The environment of the plant differs from that of red blood

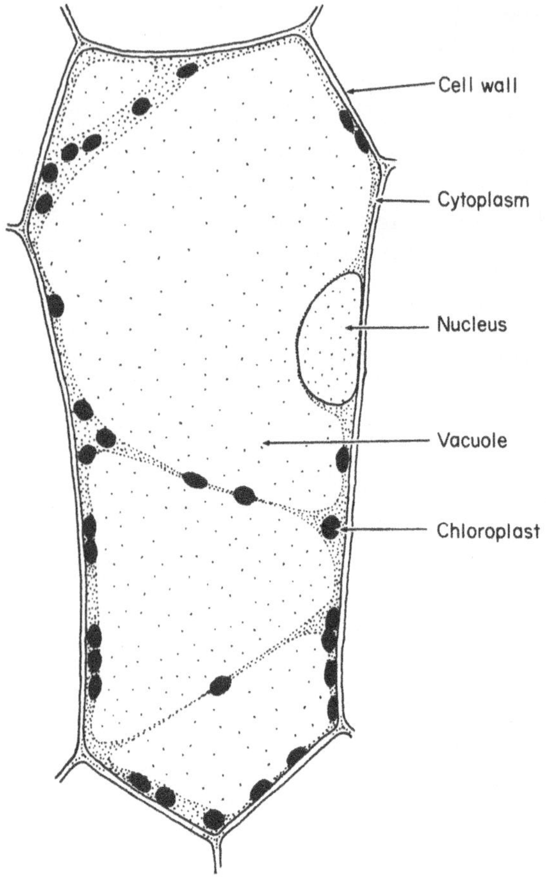

FIG. 5-1. Mature cell of *Elodea.*

cells. The blood cells occur in an environment with an osmotic concentration equal to 0.15 *M* sodium chloride. *Elodea,* however, lives in fresh water which has an osmotic concentration approximating 0.01 *M* sodium chloride.

NOTE: The osmotic concentration of fresh water is highly variable, but 0.01 *M* NaCl is a reasonable working value for our purposes.

Remove a small leaf from the growing tip of *Elodea.* Obtain as small a leaf as possible because a leaf of only one cell layer is much easier to study. Add the leaf to several drops of 0.01 *M* NaCl on a clean microscope slide and cover the preparation with a cover slip. If necessary, add more salt solution at the edge of the cover slip. This wet mount will not dry out if you will add salt solution to the edge of the cover slip whenever necessary.

Examine a cell under high power of the microscope. With the aid of the drawing (Fig. 5-1), identify the essential parts of the cell. Note especially that the central area of the cell is filled with a vacuole which presses the nucleus and cytoplasm against the cell wall.

Add 0.30 *M* NaCl at one edge of the cover slip. Draw the 0.01 *M* NaCl off from the other side with a corner of paper toweling placed against the opposite edge of the cover slip. In this way the 0.01 *M* NaCl can be replaced with the more concentrated solution. Focus on one particular cell and describe the changes that occur. The withdrawal of the cytoplasm from the cell wall is called *plasmolysis*. When the cytoplasm withdraws, the presence of the cell membrane can be inferred by the way in which the cell contents round up. If you measure the distance between the edge of the cytoplasm and the cell wall (or use some measure of your own) at regular intervals, you can measure the rate of change. Follow the change until stability is reached.

C. Presentation of Results

Prepare a written report according to the directions presented in Exercise 3. The introduction should include a discussion of the hypothesis that the cell membrane is differentially permeable and of the use of the model to test the characteristics of the membrane. The discussion should center on whether the data obtained with red blood cells and *Elodea* are interpretable as being consistent with the characteristics of a differentially permeable membrane as demonstrated by the model. The discussion should include, also, any properties of the model not tested by the experiments.

SOME FURTHER EXPERIMENTS

If time, inclination, and interest are sufficient, several additional experiments may be run. As appropriate, the results of these experiments should be incorporated in the written report.

At what maximal concentration of NaCl is there no plasmolysis in *Elodea*?

What happens to red blood cells placed in 0.01 *M* NaCl?

What does the difference between the responses of *Elodea* and red blood cells to 0.01 *M* NaCl suggest about the function of the cell wall?

Do red blood cells and/or *Elodea* respond differently if the test solutions are sucrose instead of NaCl?

Does sucrose diffuse through the cell membrane? What would this result tell you about the validity of the model?

Many freshwater organisms, such as *Paramecium* and *Amoeba*, lack a cell wall. How do these cells avoid bursting?

HINT: Fresh water cells have a contractile vacuole which is usually absent in marine species.

How could you test the hypothesis that the contractile vacuole regulates osmotic concentration of the cellular fluids?

The body fluids of freshwater animals have an osmotic concentration higher than that of the environment. What problem would this relationship create for these organisms? If their cells lack contractile vacuoles, how can they maintain osmotic equilibrium?

SUGGESTED READINGS

Guthe, Karl F. (1968). "The Physiology of Cells," 116 pp. Macmillan, New York. Paperback.

Howland, John L. (1968). "Introduction to Cell Physiology: Information and Control," 214 pp. Macmillan, New York.

EXERCISE

6 EXPERIMENTAL ANALYSIS: THE USE OF CONTROLS

The previous exercises have introduced the basic procedures of scientific investigation. Most of the work emphasized the use of quantitative data to test a hypothesis or to examine a relationship between two variables (e.g., rate of diffusion and molecular weight). In this exercise we wish to further refine the experimental approach to testing a hypothesis. Of special emphasis is the idea of using a control.

In any experiment there may be several variables acting on the system under analysis. Not all of these variables may be known to the investigator. How, then, can the investigator correct for or take into account effects of variables that may cause a misinterpretation of his results?

Let us first look at an example from drug testing. Company Q has developed a drug which it believes will cure colds. How can it be tested? Obviously, the drug must be administered to people with colds. So the company establishes a testing program and learns that 70% of the people given the drug were cured in 3 days. What a discovery! But wait! How many of those would have recovered anyway? And what about all the home remedies (like bourbon and ginger ale) that seem to effect cures? Is there a possibility that many "cures" are really natural recovery that would have occurred without the treatment? Thus another test is set up. Half the patients receive the drug and the other half receive a placebo (a medication identical to the drug in appearance, taste, etc., but which has no value other than psychological). Again, 70% of the people receiving the drug are cured, but so are 50% of the people receiving the placebo. Thus, the drug is only 20% effective and not worth marketing. In this experiment the placebo was the control. The control provided a basis for separating the effects of the drug from those effects which could not be measured.

Thus, a control in any experiment is used to provide a basis for making a valid comparison. The control serves to indicate the amount of response that occurs due to variables not anticipated. In a classic experiment, all factors are held constant or controlled except the one factor whose effects are under investigation. The difference between the controlled results and the experimental results is the effect which may be attributed to the experimental manipulation.

1. USE OF CONTROLS IN THE STUDY OF RESPIRATION

Respiration may be defined as the slow, controlled release of energy through enzymatic activity. Respiration may be aerobic or anaerobic. The input-output equations for each reaction may be written as follows:

Aerobic:

$$\underset{\text{glucose}}{C_6H_{12}O_6} + 6\ \underset{\text{water}}{H_2O} + 6\ \underset{\text{oxygen}}{O_2} \xrightarrow{\text{enzymes}} 6\ \underset{\text{carbon dioxide}}{CO_2} + 12\ \underset{\text{water}}{H_2O} + \underset{\text{energy (456,000 calories per mole)}}{e}$$

Anaerobic:

$$\underset{\text{glucose}}{C_6H_{12}O_6} \xrightarrow{\text{enzymes}} 2\ \underset{\text{carbon dioxide}}{CO_2} + 2\ \underset{\text{ethyl alcohol}}{C_2H_5OH} + \underset{\text{energy (24,000 calories per mole)}}{e}$$

$$\underset{\text{glucose}}{C_6H_{12}O_6} \xrightarrow{\text{enzymes}} 2\ \underset{\text{lactic acid}}{C_3H_6O_3} + \underset{\text{energy (36,000 calories per mole)}}{e}$$

The energy in the above input-output statements is incorporated into high-energy phosphate bonds which are used to form adenosine triphosphate.

Anaerobic respiration may lead to ethyl alcohol or to lactic acid. The process leading to ethyl alcohol classically was called fermentation; the process leading to lactic acid was known as glycolysis. However, not all current usage of fermentation and glycolysis follows the classic pattern. The energy yield of fermentation (which occurs in certain Protista such as yeast) is less than that of glycolysis (which occurs in animals and certain bacteria).

A. Cellular Respiration in Yeast

Yeast is a one-celled fungus that is capable of respiring aerobically or anaerobically by fermentation. The yeast obtains about 3800 calories/gm of glucose aerobically and about 140 calories/gm of glucose by fermentation. If oxygen is available, a given batch of yeast cells will respire aerobically and anaerobically at the same time. As the availability of oxygen changes, one process or the other will be accentuated. For the most part, respiratory processes in yeast are identical with those of any other organism. As yeast is readily handled in the laboratory, we shall use it in the initial experiments to demonstrate the use of controls.

Many of the biochemical steps in the transformation of glucose to ethyl alcohol require the presence of magnesium. Magnesium ions activate several enzymes in this series of chemical events. The experiment will investigate the role of magnesium in affecting the rate of respiration. Additional Mg may be added as a soluble salt; on the other hand, Mg may be removed by precipitation as a fluoride. Glucose is the substrate which is oxidized.

One final problem remains. How is respiration measured? In fermentation, the oxidation of a mole of glucose produces 2 moles of carbon dioxide and 2 moles of ethyl alcohol. Thus one might measure any of these three because they are directly proportional to one another. That is, the decrease in glucose, the increase of carbon dioxide or of ethyl alcohol can be used as a measure of respiration. Because carbon dioxide is emitted as a gas and can be easily collected, it will be measured in the following experiments.

1. *Experimental procedure.* First, two precautions: Be certain that all glassware is *thoroughly* clean. If necessary, scrub with soap and water, rinse several times with tap water, drain, and rinse twice with distilled water. Fluorides are lethal *poisons.* The concentrations you are using are far below the lethal level. However, wash your hands after handling solutions of fluoride. Rinse your mouth thoroughly if you should suck up any fluoride into it. Clean glassware painstakingly after use. Be *cautious* and you will have no difficulty.

Respiration may be measured with a simple test tube–vial respirometer. Practice filling and assembly of a respirometer according to the following directions:

1. Fill the test tube with distilled water.
2. Hold a vial upside down and slide it over the test tube so that the bottom of the vial comes to rest against the open rim of the test tube.
3. Using your fingers if necessary to hold the tube and vial in position, quickly invert the entire system.
4. The vial will now be right side up and the test tube will be upside down in the vial. There will be an air space over the water column in the test tube. The size of the air space will depend on your technique and should be kept as small as possible.
5. Measure with a millimeter ruler the length of the air space from the bottom of the meniscus to the highest point of the test tube.

Dismantle the practice run, empty the distilled water, and number the test tubes from 1 to 5 (hold the tube upside down as you number). Place the test tubes in the rack and proceed with the following.

Add to each of the tubes the amounts of each solution as indicated by the tube number. Each test tube will hold 15 ml of fluid. You may use your time more economically if you add the yeast suspension to each tube, then add the glucose, then the distilled water, etc. Tube 1 is the control. Any gas production in this tube occurs because of substrate, Mg, etc., already present in the cells. Also, this tube corrects for gas production that may arise from unexpected sources. Tube 2 is a standard. Substrate only is supplied. Respiration occurs because the yeast cells contain some magnesium. The standard provides the base data against which experimental manipulations may be compared.

Tube number 1: 5 ml yeast suspension, 10 ml distilled water
2: 5 ml yeast suspension, 5 ml glucose, 5 ml distilled water
3: 5 ml yeast suspension, 5 ml glucose, 5 ml 0.10 *M* $MgSO_4$
4: 5 ml yeast suspension, 5 ml glucose, 5 ml 0.06 *M* NaF
5: 5 ml yeast suspension, 5 ml glucose, 5 ml 0.20 *M* NaF

NOTE: The test tubes hold exactly 15 ml. After the first 10 ml are added, the remaining 5 ml may be added by pouring to fill the tube.

Now slide the vials over the filled small test tubes and invert, as outlined above. Place the respirometers in the test tube rack.

2. *Recording data.* In the appropriate space in Table I, record the initial gas reading of each tube in milliliters. To do this, measure the height of the air space with your millimeter rule and convert this to milliliters by assuming that every 7-mm length of air column in the test tube corresponds to an air volume of 1 ml. Remember that this length-to-volume relationship applies only to test tubes of this diameter. If a test tube of different diameter is used, the volume may be calculated by using the formula for the volume of a cylinder: $V = \pi r^2 h$.

TABLE I

	1	2	3	4	5
Contents	10 ml H_2O	5 ml H_2O	0.10 *M* $MgSO_4$	0.06 *M* NaF	0.20 *M* NaF
Initial gas reading (ml)					
Terminal gas reading (ml)					
Net gas volume (ml/hour)					

If you are a bit nervous, or have five thumbs on each hand, you might jiggle the tubes and allow air to enter. If so, you must be sure that the initial gas reading is corrected for this error.

Now place your respirometers into the water bath of 37° C. After 1 hour, remove the respirometers and in Table I record the terminal gas reading. Subtract the initial from the terminal readings, and record the net gas volumes per hour.

Clean all glassware before leaving the laboratory.

3. *Discussion of results.* The following hypothesis may be stated: Magnesium is essential for respiration in yeast cells. Use data obtained in this experiment to support or reject this hypothesis. Discuss fully.

B. Effect of Substrate on Rate of Respiration

If we wished to analyze respiration further, there are many additional questions that might be asked. For example, does temperature or light affect respiration? What effect does ethyl alcohol have? Does the nature of the substrate affect the rate of respiration? Let us examine this latter question further.

In the previous experiment, the simple sugar glucose was the substrate. Such a six-carbon sugar is a monosaccharide. *Fructose,* also called levulose or fruit sugar, and *galactose* also are monosaccharides. Fructose and galactose have the same empirical formula as glucose, but the structural pattern or arrangement of the atoms in the molecules differ. Therefore, glucose, fructose, and galactose might not be oxidized at the same rates because they might react differently with the enzymes that control the rates of the varied steps of cellular respiration.

Only the simple sugars proceed directly into respiration. Double sugars, or disaccharides, and multiple sugars, or polysaccharides, must be digested into simple sugars before undergoing respiration. Digestion requires enzymes specific for the particular substrate. Three disaccharides are *sucrose, maltose,* and *lactose.* Sucrose consists of one molecule of glucose and one molecule of fructose. Maltose consists solely of glucose. Lactose has one molecule of glucose and one molecule of galactose. The polysaccharide *starch*, a characteristic storage product of plants, is composed of units of glucose.

Prepare eight numbered respirometers. Add 5 ml yeast suspension and 5 ml distilled water to each tube. Then add the following:

Tube number 1: 5 ml starch
2: 5 ml lactose
3: 5 ml sucrose
4: 5 ml maltose
5: 5 ml fructose
6: 5 ml galactose
7: 5 ml glucose
8: control: an additional 5 ml distilled water

Why is a tube containing glucose included?

Record the initial gas reading in the Table II and place your respirometers in the water bath. After an hour, remove your respirometers and complete the table.

TABLE II

Tube no.	1	2	3	4	5	6	7	8
Contents	Starch	Lactose	Sucrose	Maltose	Fructose	Galactose	Glucose	Control
Initial gas reading (ml)								
Terminal gas reading (ml)								
Net gas volume (ml/hour)								

Evaluate and discuss your results. Perhaps you can formulate one or more hypotheses concerning differences in rates of respiration by yeast cells supplied with different sources of energy.

2. OXYGEN UPTAKE IN MULTICELLULAR ORGANISMS

In Part 1, cellular respiration was studied by measuring carbon dioxide production. In aerobic respiration, oxygen is utilized. Thus respiration may be measured by determining the amount of oxygen used. The vast majority of multicellular organisms are aerobic and the determination of oxygen uptake is a standard way of studying respiration.

If you will look again at the input-output statement or equation for aerobic respiration, you will note that the amount of carbon dioxide produced is equal to the amount of oxygen consumed. That is, for every mole of oxygen used, a mole of carbon dioxide is liberated. Thus respiration could be measured in aerobic organisms by determining the amount of carbon dioxide produced. However, generally it is easier to measure oxygen uptake.

There is one further complication to using carbon dioxide production as a measure of respiration which should be considered. Carbon dioxide output equals oxygen uptake only when carbohydrate is the substrate. When fat or protein forms the substrate, more oxygen is used than carbon dioxide produced. The following input-output equations demonstrate this relationship:

$$\underset{\text{fat}}{2\ C_{51}H_{98}O_6} + \underset{\text{oxygen}}{145\ O_2} \xrightarrow{\text{enzymes}} \underset{\text{water}}{98\ H_2O} + \underset{\text{carbon dioxide}}{102\ CO_2} \qquad (1)$$

$$\underset{\text{protein}}{C_{492}H_{728}O_{59}N_{120}} + \underset{\text{oxygen}}{615\ O_2} \xrightarrow{\text{enzymes}} \underset{\text{water}}{364\ H_2O} + \underset{\text{carbon dioxide}}{492\ CO_2} \qquad (2)$$

The relationship of ratio of carbon dioxide output to oxygen uptake is called the *respiratory quotient* (RQ). The ratio may be expressed as:

$$RQ = CO_2/O_2$$

The RQ varies with the substrate. Thus, if carbohydrate is the substrate, the $RQ = 6\ CO_2/6\ O_2 = 1$. From Equation (1), the RQ for fat $= 102\ CO_2/$

145 O_2 = 0.7. From Equation (2), the RQ for protein = 492 CO_2/615 O_2 = 0.8. You can easily see that each substrate has a characteristic RQ. By measuring the RQ, some idea of the type of substrate being utilized can be determined. Of course, you should keep in mind that at any one time various substrates are generally utilized by an organism, but one or the other substrate may predominate under certain conditions, such as starvation, periods of food storage, etc.

If, in the following experiments, RQ is determined, both carbon dioxide and oxygen must be measured. Oxygen consumption may be measured readily with a jar respirometer. There are two ways in which the jar respirometer may be used. The jar may be set upright and stoppered with a one-hole rubber stopper. A piece of glass tubing is bent at a right angle. One arm is inserted in the rubber stopper and the other arm is horizontal to the table. An indicator fluid is placed in the horizontal arm. If the respirometer is laid on its side, a straight piece of glass tubing can be used. For this setup, a calibrated pipette makes a fine respirometer tube and greatly simplifies calculation of oxygen consumption. In either setup, the respirometer tube (the arm or tube with the indicator fluid) must be horizontal. For small organisms, shell vials may be used instead of jars. One further point. The respirometer tube must be clean.

Let us emphasize some sources of error which you must control if you are to obtain accurate results. These simple respirometers operate on the principle that as oxygen is consumed in the jar, the air pressure is lowered and the indicator fluid in the respirometer tube moves inward. Thus the stopper must be airtight to prevent leaking in of air which would raise the pressure inside the jar and prevent the indicator fluid from moving or cause it to move in the wrong direction.

Because carbon dioxide is produced by the organism, this gas will accumulate inside the jar and form a back pressure against the indicator fluid so that its movement will not be sufficient to reflect the amount of oxygen actually consumed. This source of error may be eliminated by absorbing the carbon dioxide with potassium hydroxide or soda lime.

The measurement of carbon dioxide is more difficult, but may be approximated by measuring oxygen utilization with a carbon dioxide absorbent and repeating the measurement without the absorbent. The difference in oxygen uptake between the two determinations is a measure of the carbon dioxide produced. If you should attempt this measurement, think of other sources of error and write them down so that you can take steps to avoid them.

Change of temperature may be a marked source of error. An increase in temperature will expand the gases inside the respriometer; a decrease in temperature will contract the gases. What effect will expansion or contraction of the gases have on the movement of the indicator fluid in the respirometer tube?

The amount of respiration which occurs must be expressed per unit of time. Why is the expression of respiration as a rate necessary?

The calculation of the amount of oxygen consumed is determined by measuring the volume of a cylinder. The volume of a cylinder is calculated from the formula $V = \pi r^2 h$. In this formula, h is the distance the indicator fluid moves along the respirometer tube. If a tube of 2 mm diameter is used, the radius (r) is one and $r^2 = 1$. Thus the volume of the cylinder is easily calculated as π (3.14) $\times$ h (the distance in millimeters the indicator fluid moved) = cubic millimeter of oxygen consumed.

CAUTION: Check the diameter of the respirometer tube! If the diameter is not 2 mm, you must incorporate the r^2 into your calculation. The final step of the calculation is the division of the amount of oxygen consumed by the appropriate time factor to give the rate of oxygen consumed per minute or per hour. You may wish to express the rate of oxygen consumption per minute or per hour, depending on the magnitude of the numbers, but use the same time scale throughout your experiments so that the data are readily comparable. The calculation is summed (for a tube of 2 mm diameter):

$$\frac{\text{mm}^3\text{O}_2\text{ consumed}}{\text{min}} = \frac{3.14 \times \text{mm indicator fluid moved}}{\text{time (min)}}$$

A. Investigation of a Problem in Oxygen Consumption

Each student or group of students choose a simple problem to investigate. A number of suggestions are given below, but the student should feel free to investigate a problem which reflects his own idea.

State the problem as a hypothesis. For example, a hypothesis might be: Oxygen consumption of seedlings is independent of size of seedlings.

Carefully outline your methods, including the controls, and list the materials you will need. Construct a table or devise some other way of recording the data.

Finally, write a short report of your experiment following the procedures used in previous exercises. The discussion should concentrate on whether the data support the hypothesis.

B. Suggested Problems

1. *Seedlings.* Seedlings are used to measure oxygen consumption because they are small enough to handle readily in the laboratory respirometers and because seedlings are active organisms. There are several problems that could be chosen for investigation:

   a. Does light affect the rate of oxygen utilization?
   b. How does temperature affect the rate of oxygen consumption?
   c. What substrate is the seedling utilizing?
   d. Do different kinds of seedlings respire at the same rate?

2. *The mouse.* Some of the same questions that were asked of the seedling may also be asked of the mouse. Some other questions are:

   a. Do different strains of mice respire at different rates?
   b. Does a well-fed mouse use the same substrate in respiration as a starved mouse?

The work may be expanded to include other animals. For example, the American chameleon, *Anolis carolinensis,* is easily kept in the laboratory and is a poikilotherm whose responses could be compared to the homeothermic mouse. If available, insects, such as cockroaches, make good subjects. If a variety of kinds of organisms are present, you may wish to direct your problem to comparing how different species respond to some environmental factor.

EXERCISE

7 EXPERIMENTAL ANALYSIS: *IN VITRO*

In Exercise 6, the relationship of magnesium to respiration was examined in living yeast cells but we did not determine what enzymes were affected or if the enzymes were affected to the same degree by the addition of or removal of magnesium. How can an individual reaction which occurs in cells be analyzed independently of all the other reactions taking place in the same cells? The discovery late in the nineteenth century that living processes could occur outside the cell opened up the possibility of analyzing cellular reactions and made possible the great advances of molecular biology in this century. The occurrence of living processes in a nonliving environment is designated *in vitro*. *In vitro* literally means in glass. By contrast, *in vivo* refers to the processes occurring in living organisms.

Cautions must be exercised in interpreting results determined *in vitro*. The environment in which the reaction occurs undergoes drastic change from the living cell to the test tube. Thus specific results from an *in vitro* analysis may often serve as a model or give only an indication as to the "true" role inside the cell. Nevertheless, if *in vitro* results are interpreted sensibly, they can contribute much to our understanding of living processes.

One of the best examples of *in vitro* analysis is the investigation of enzyme function. Enzymes are proteinaceous substances that regulate reactions without undergoing marked change in their structure.

1. EXAMINATION OF AMYLASE *IN VITRO*

Amylase is an enzyme which catalyzes the hydrolysis of the polysaccharides starch and glycogen. The first products of hydrolysis are various intermediates known as dextrins (polysaccharides of various chain lengths). The dextrins are further broken down into a double sugar, maltose, and a small amount of glucose.

The enzyme has been prepared in physiological saline for your use so that 4–8 drops will digest 5 ml of 1% starch solution in about 10 to 15 min. Because the strength of the enzyme may vary from one preparation to the next, your instructor will tell you how many drops to use. The enzyme solution will be referred to as *standard enzyme solution* in subsequent directions. The presence of starch may be determined by adding a drop of iodine solution to a drop of starch solution. If starch is present, a blue to a deep blue-black color develops. As digestion proceeds and the starch is hydrolyzed into dextrins, the color changes from blue-black to reddish-violet to reddish-brown. When digestion is completed, only the brownish color of iodine is present. The point at which the reddish-brown changes to the iodine color is called the *achromic point*. That is, there is no further color reaction between the iodine and the substrate (starch and dextrins). In practice, it is not necessary for the reaction to go to completion. Because the rate of enzyme activity will be investigated, it is important that sufficient time elapse so that any differences in rate of enzyme activity under different experimental conditions can be detected. Therefore, it is more critical that an *endpoint* of the reaction be chosen and that the same endpoint be used throughout an experiment. The endpoint should coincide with one of the color changes described above, e.g., when the color changes to reddish-violet or reddish-brown. Once the endpoint or achromic point is reached, further testing for the presence of starch is unnecessary. One further point. The achromic point and the endpoint are not necessarily the same. Ideally, the achromic point is a fixed point and occurs when there is no further color response. The endpoint may arbitrarily be chosen at some convenient spot by the investigator, but should be one that can be repeatedly recognized. In these experiments we shall use the designation achromic point. It is advisable to make a test run in order to observe the nature of the color change and to choose an endpoint for use in all subsequent experiments.

In each of the following experiments, several test tubes of enzyme and substrate will be run concurrently. Thus several tests must be made in rapid order. The testing may be performed on porcelain spot plates or on glass plates laid on white paper. Whichever method you use, label the spot plate or paper so that you know which starch solution was added to which drop of iodine.

All testing for starch will use the following procedure:

1. Place 1 drop of iodine solution in each of several depressions on a porcelain spot plate or spaced on glass plates. Use the same number of drops of iodine solution as the number of tubes of starch solution. Iodine solution deteriorates upon exposure to air. Do not permit drops to stand more than 2 min before use.
2. Add a few drops of starch solution to a drop of iodine. Use each drop

only once. Do not add drops of starch solution from two different test tubes to the same drop of iodine.

3. Begin testing no more than 1 min after adding enzyme to the starch solution.
4. Repeat the above procedure each minute.

The following experiment will be performed by students working in pairs:

A. Effect of pH on Enzyme Activity

Set up a duplicate series of five tubes numbered 1 through 5. Place the tubes in two rows so that tube 1 is opposite its duplicate, etc. Place 5 ml of 1% starch solution in each tube. The 1% starch solution is made up in buffers so that each starch solution has a known pH. Arrange your solutions to conform to the following pattern:

Tube number 1: 5 ml of 1% starch solution, pH 4
(includes 2: 5 ml of 1% starch solution, pH 6
duplicate) 3: 5 ml of 1% starch solution, pH 7
4: 5 ml of 1% starch solution, pH 8
5: 5 ml of 1% starch solution, pH 10

Record the temperature at which the solutions are maintained. Frequently temperature is recorded as room temperature. Room temperature is assumed to be 21°–22° C.

Add the proper number of drops of *standard enzyme solution* to each tube in one row and add the same number of drops of distilled water to each tube in the second row. Mix well. The tubes in the second row serve as controls. How often is it necessary to test the controls?

Follow digestion by means of the iodine test (measure at 1-mm intervals) and determine the time required for each tube to reach its achromic point (the time when the starch test is negative). Then determine the reciprocal of the time ($1/T$) required for digestion for each pH. Enter your results in Table I.

TABLE I

Tube number	pH	*T*-Achromic point (time for digestion)	$1/T$
1	4		
2	6		
3	7		
4	8		
5	10		

The reciprocal of time, $1/T$, is used for two reasons. First, the reciprocal is more narrowly distributed than time; hence the values are less widely ranged when graphed. Second, the reciprocal inverts the curve so that the high rates of enzyme activity plot as a peak on the graph paper rather than as a valley. You may easily demonstrate this by plotting a set of data as T and as $1/T$.

The following hypothesis may be stated: The rate of digestive activity of amylase is independent of the pH of the environment in which the enzyme functions.

Plot the data as $1/T$ for each pH against pH on the following graph:

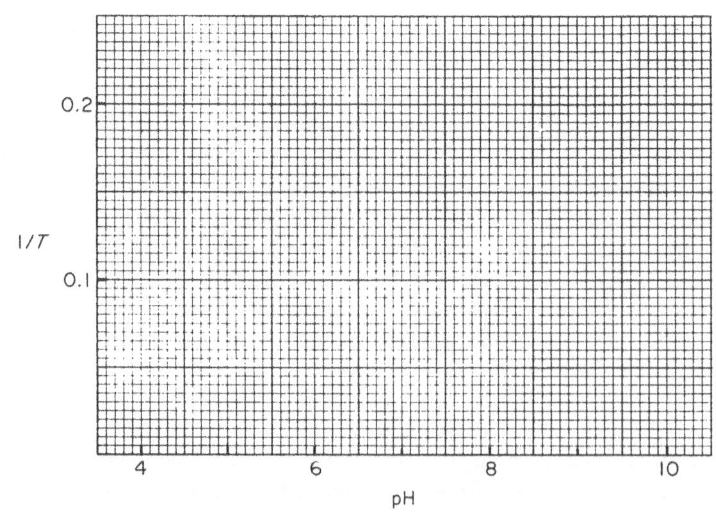

Discuss whether the hypothesis is accepted or rejected.

B. Independent Problems

Each student or pair of students will investigate the effect of some factor on the rate of activity of amylase. The methods and materials may be essentially those described in Part A.

The following problems are presented as suggestions for possible study. The problems are stated as general hypotheses which might be made more specific for a particular problem.

1. The activity of amylase is independent of temperature.
2. The activity of amylase is independent of the concentration of the enzyme.
3. The activity of amylase is independent of substrate concentration.
4. The activity of amylase is not affected by ions.

Discuss your problem with your instructor. The instructor will approve the feasibility of investigating the problem, but you are responsible for its design.

Prepare a written report, following the procedures outlined in Exercise 3, which describes your investigation of the relationship of some variable to enzymatic activity.

SUGGESTED READINGS

Baker, J. and Allen, G. (1965). "Matter, Energy, and Life," 180 pp. Addison-Wesley, Reading, Massachusetts. Paperback.

Coult, D. A. (1966). "Molecules and Cells," 166 pp. Houghton Mifflin, Boston, Massachusetts. Paperback.

Jellinck, P. H. (1967). "The Cellular Role of Macromolecules," 117 pp. Scott, Foresman, Chicago, Illinois. Paperback.

EXERCISE

8 FUNCTIONAL ORGANIZATION OF FLOWERING PLANTS

In the first five exercises in this manual, the process of investigation was emphasized. In a subsequent series of exercises, the process of experimental analysis will be examined further. Some of these exercises and others in the latter half of this manual require some knowledge of plant and animal organization. Therefore, at this time, attention will focus on the functional organization of multicellular organisms.

The single, most significant problem of all organisms is obtaining energy. Energy comes from the environment. Therefore, much of plant and animal organization can be analyzed as structural devices for extracting energy sources from the environment.

The second most significant problem of all organisms is coping with the physical, chemical, and biological environment. No organism can ignore gravity, or the seasonal changes of temperature and moisture. Do not assume that organisms are conscious of these problems in the sense that we are. Rather, consider that natural selection favors those patterns of organization that have the higher probability of survival. Those patterns of organization, both structural and functional, that combine relatively efficient means of obtaining energy with devices for withstanding environmental stress have the advantage over those patterns which are less effective in satisfying these two demands. We will not have time to examine the myriad variations of the theme of adaptation (ability to cope with the environment) or to explore the varied means of achieving the same end. But as you examine plant and animal organization, a consideration of how its organization prepares the organism to exist in its environment and obtain energy will be the most fruitful aspect of your analysis.

Energy transformations occur in cells. These transformations require materials and usually result in the formation of waste products. The cell obtains its energy source from its environment and eliminates its metabolic wastes into the environment. These activities, then, result in an exchange between a cell and its environment.

For simpler plants and animals, the environment of a single cell is also the environment of the whole animal. Metabolism is maintained primarily by exchange between the cells and the water of the external environment. Thus the cells are affected by any change in the environment. Because any cell has a narrow range of environment over which it can live, such organisms can live only in those environments that do not fluctuate too widely. The great majority live in the sea, although a few live in fresh water.

In the course of evolution, organisms have achieved varying degrees of independence from their environment. This independence is made possible by enclosing within the whole organism an internal environment and shifting the fundamental activities of maintenance to the internal cells. Thus the cells operate in a stabilized environment. The external environment of the cells, in a sense, is the internal environment of the whole organism.

A complex organization of systems is needed to maintain the constancy of the internal environment of an organism. Food, oxygen, and nutrients must be supplied to the cells; metabolic wastes must be carried away from the cells. In addition, these substances must be exchanged between the internal environment of the organism and its external environment.

Multicellular organisms are too large to enable diffusion to serve as the means by which individual cells are supplied with needed materials and as the means by which waste products are eliminated. Therefore, all of these organisms have some type of circulatory system. (There are exceptions, some multicellular animals are quite small and do not have circulatory systems. Also, some plants do not have special transport tissues.) In summary, the maintenance of an organism requires three essential functions: obtaining energy, transport, and exchange.

1. ORGAN SYSTEMS OF A PLANT

Our discussion of the organization of the flowering plant will include those organs involved in maintenance. Reproduction is not an essential activity of an organism. Indeed, it is immaterial to the organism whether or not it reproduces. Reproduction is essential to the population and a population cannot be maintained without at least some individuals reproducing. Because populations are composed of units called organisms (or individuals), organisms reproduce. The reproductive organ of the flowering plant is, of course, the flower.

If you are not already familiar with the general morphology of a plant, obtain a young bean or pea plant and identify the roots, stem, and leaves. Among the essential functions, the root is specialized for exchange; the leaf, for transforming energy and exchange, and the stem, for transport. Additionally, the root and stem provide support and may serve as organs of storage of chemical energy, primarily in the form of starch.

The discussion of the functions of root, stem, and leaf emphasizes another characteristic of biological systems: specialization. Specialization occurs at all levels of biological organization. Within the cell the mitochondrion is specialized for carrying on certain biochemical steps in respiration; the chloroplast, for photosynthesis. Cells may be specialized for support, storage, transport, synthesis, contraction, or protection. Likewise, tissues and organs may be specialized for given functions. In these more complex structures, the overall function is achieved by the integration of populations of specialized cells, such as those of an organ.

As you examine the detailed organization of each of the three major organs of maintenance, keep in mind the general concepts of specialization and organization discussed above.

A. The Functional Organization of the Stem

An organ is a part of an organism which is specialized for a particular function and which is composed of two or more tissues. A tissue is a population of cells of a particular type which usually has a single major function.

All specialized cells of a mature organism are derived during development from a more generalized type. The generalized plant cell which is capable of dividing is the meristematic cell. The cell types that we will observe in the stem are the result of complex processes of differentiation. If we could see the ultrastructure of these cells, more differences would become evident. However, we will work at the level of the light microscope because the principles of cell specialization or organ organization are evident at that level of magnification.

With the aid of the illustration, examine the cross section of stem of *Medicago* (alfalfa) (Fig. 8-1). Four kinds of cells, classified on a single functional basis, may be found.

Parenchymatous cells are usually large and thin-walled. They are relatively undifferentiated, maintain their living contents, and form the bulk of primary tissues in plants. Primary tissues are those that are derived from apical meristems. Secondary tissues are those which are derived from the cambium (to be described later). Primary growth is chiefly responsible for elongation of roots and stems; secondary growth is mainly responsible for increase in thickness. We will not consider secondary growth in this exercise. Parenchyma may be photosynthetic when it contains chloroplasts or may be adapted for storage of food and water.

The center of the stem is filled with parenchyma. What would you judge its function to be? Do you ever eat the stems of plants? How can you determine that the central cells are not photosynthetic?

Other parenchymatous cells are located in several layers around the circumference of the stem. Some of these cells contain chloroplasts.

Protective cells form the outer layer of cells encircling the stem. These cells

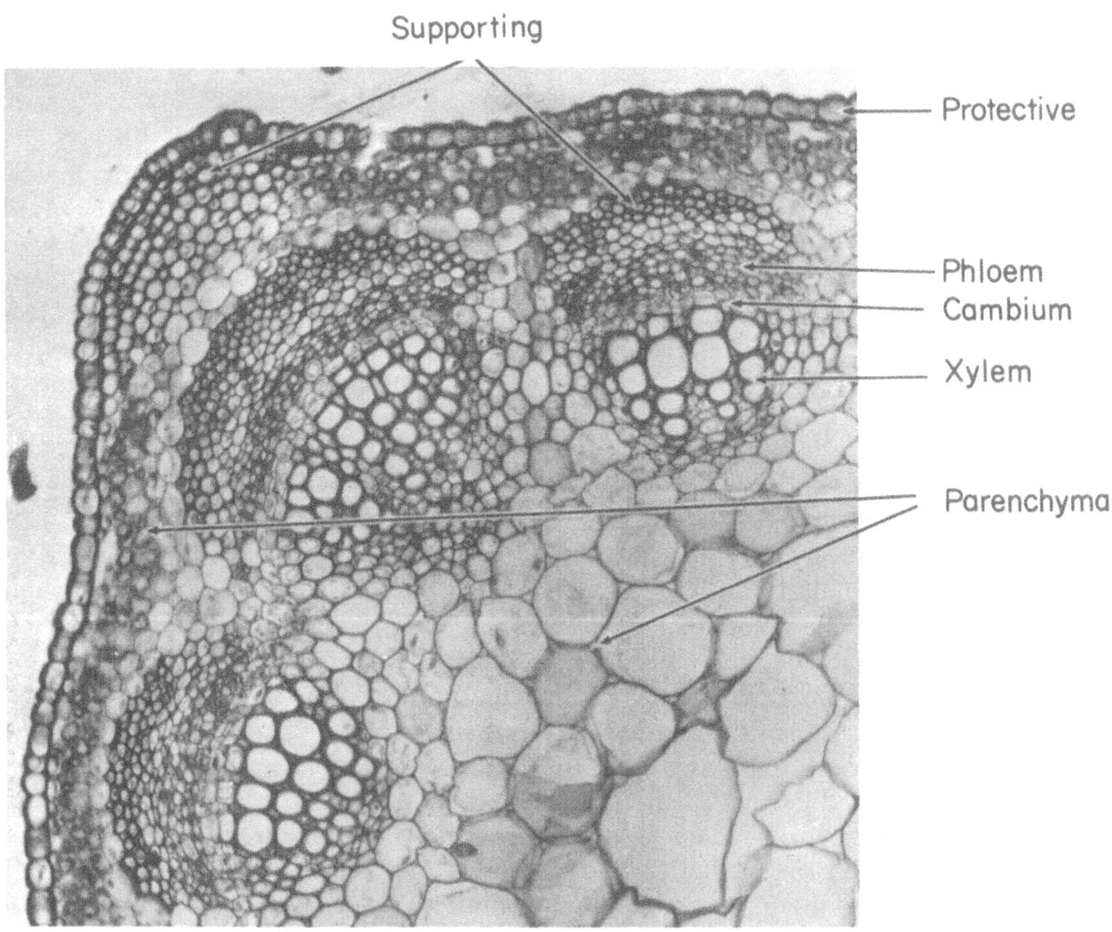

FIG. 8-1. A quarter of a cross section of the stem of *Medicago*.

make up the epidermis and are covered with a layer of water-impervious substance which forms a cuticle. The cuticle is secreted by the protective cells.

Meristematic cells in the stem form a single or double layer of elongate cells called the *cambium*. Division of cambial cells results in secondary thickening of the stem.

Supporting and conducting cells occupy large areas of the stem. Conducting cells are of two main types. The *xylem* contains cells which are thick-walled with a relatively large central space or lumen. Water and mineral ions are conducted in these cells. The cells consist only of cell walls at maturity. The *phloem* consists of cells which have relatively thin walls and contain cytoplasm at maturity. Phloem cells transport a variety of materials such as sugars and probably amino acids and mineral ions. Both xylem and phloem cells may be further classified into structural types, but we shall not be concerned with this terminology here.

There are various types of supporting cells, but all are characterized by the presence of thick walls. Notice the location of these supporting cells. Clumps of these thick-walled cells are found outside of the phloem. Try to visualize these cell populations forming a support column, like a rectangle of bricks, running the length of the stem. Notice also, the concentration of supporting cells at the four "corners" of the stem. These cells also form support columns. A layer of supportive cells occurs just beneath the epidermis.

Do you think that xylem cells are also supporting? On what do you base your answer?

B. The Functional Organization of the Leaf

Examine, with the aid of Fig. 8-2, the cross section of *Ligustrum*, the privet, or some other leaf. You should first locate the epidermis. The leaf has both an upper and lower epidermis. Can you see any difference in the types of cells in the two layers of epidermis?

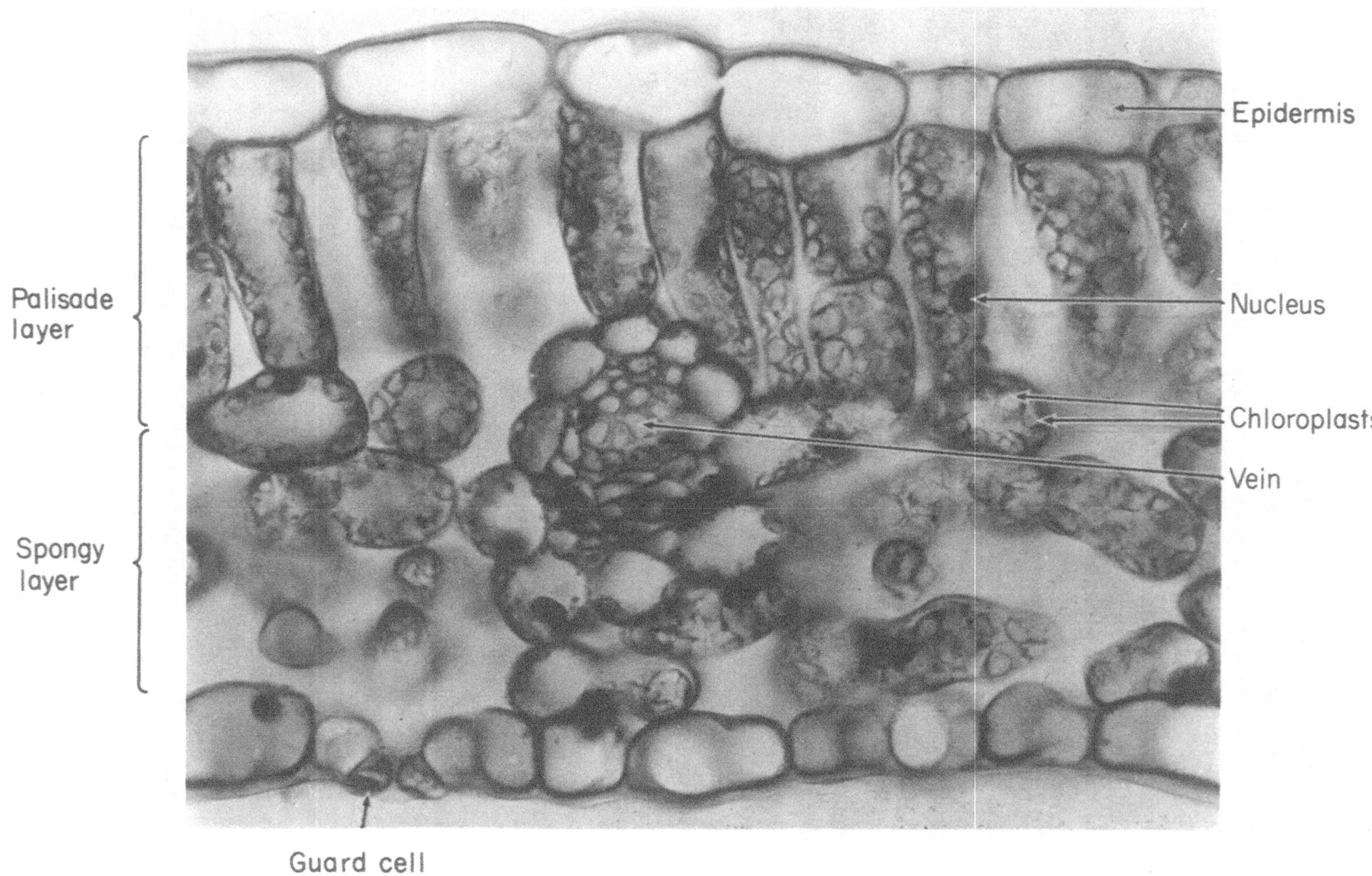

FIG. 8-2. Cross section of the leaf of *Ligustrum*.

What functional reasons can you give for the concentration of stomata on the lower epidermis?

What type of cell makes up most of the leaf?

Can you think of any reason why cells should be elongated to form the palisade layer?

Beneath the palisade layer is the spongy layer, characterized by large intercellular spaces. What advantage is there to the leaf to having such a structural organization?

In which cells would most of the photosynthesis occur? Explain your answer.

Is there any evidence for the presence of supporting and/or conducting cells? What is a vein? Have you ever observed the pattern of veins in a leaf?

Does the leaf contain cambium? How do you account for the lack of cambium in the functional organization of the leaf?

Would you expect the leaf or stem to carry on more photosynthesis? Explain your answer.

C. The Functional Organization of the Root

With the aid of Fig. 8-3, examine the cross section of the root of *Ranunculus* (buttercup). In the center of the section is a core of xylem with four ridges radiating outward. At the outer extremities of the xylem ridges are one or two to three layers of thin-walled cells. These cells form a ring known as the *pericycle*. Branch roots originate from the pericycle and the root increases in diameter as a result of the pericycle's becoming meristematic.

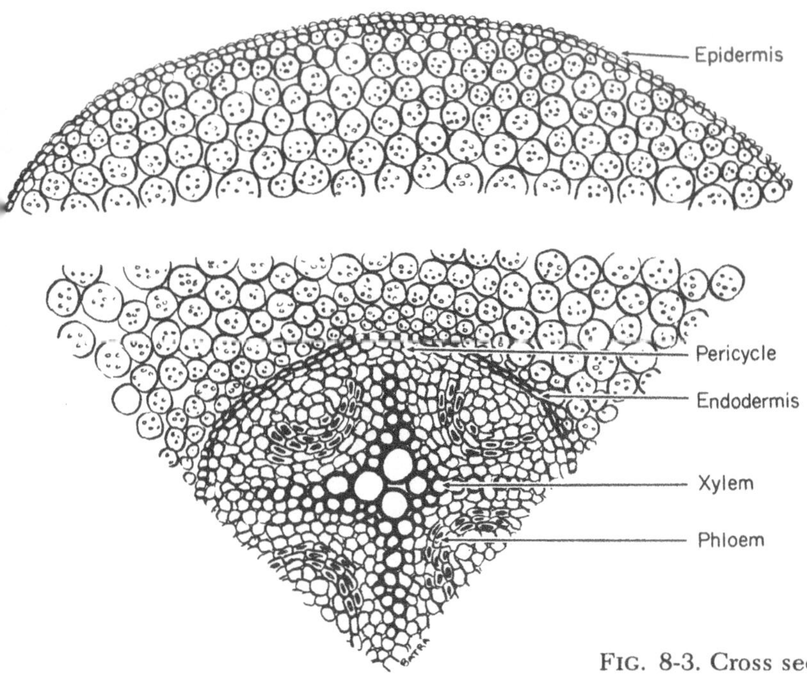

Fig. 8-3. Cross section of the root of *Ranunculus*.

Alternating with the xylem ridges and also in contact with the pericycle are small groups of thin-walled cells, the *phloem*. A parenchyma fills in the angles between xylem and phloem. Why should the root have a phloem?

Outside of and directly in contact with the pericycle is a cylinder of cells, one cell in thickness, the *endodermis*. Sone of the cells are thick-walled and others are thin-walled. The thin-walled cells are sometimes called *passage cells*. The endodermis apparently regulates the movement of materials into and out of the center of the root.

Outside of the endodermis is a broad cylinder of thin-walled parenchyma that makes up the bulk of the root. Numerous starch grains are present in the cells. What is the principal function of this parenchyma?

The outer layer of the root is the epidermis. The most important function of the root epidermis is the uptake of water. Examine the demonstration of young roots and note the presence of root hairs. Also examine the demonstration of a cross section through the root hair zone and note the relationship of the root hair to the epidermal cells. On the basis of these morphological observations, discuss the possible significance of root hairs.

SUGGESTED READINGS

Galston, A. W. (1964). "The Life of the Green Plant," 118 pp. Prentice-Hall, Englewood Cliffs, New Jersey. Paperback.

Ray, P. M. (1963). "The Living Plant," 127 pp. Holt, New York. Paperback.

Salisbury, F. B. and Parke, R. V. (1964). "Vascular Plants: Form and Function," 184 pp. Wadsworth, Belmont, California. Paperback.

EXERCISE

9 SOME INVESTIGATIONS OF PLANT ORGANIZATION

This exercise is divided into two parts. The first part consists of the morphological description of plant organs not previously discussed and the second part suggests some experiments on water transport. Each student should conduct an investigation from either group. If a lengthy experiment is set up, several students may work together.

1. MORPHOLOGICAL UNKNOWN

The principles of plant organization learned in the previous exercise may be applied to the examination of unknown material. From the previous exercise you learned the general characteristics of the various cell types and how these cell types are arranged in roots, stems, and leaves. Any of these organs may be modified for certain functions or as adaptations to particular environmental conditions. Also, the cell types might be arranged somewhat differently in the same organ from plants of different taxonomic status. Your task will be to decide whether you have a root, stem, or leaf and to describe its organization. You will also be expected to make deductions concerning the functional significance of the organization. These deductions should lead to a hypothesis about the environment of the plant, etc. Any deductions that are made should be justified. You may also wish to sketch portions of your material to illustrate your description and support your deductions. Finally, you should discuss your analysis with your laboratory instructor. If you make errors, the two of you should discuss the material so that you can correct your mistakes and continue your analysis. Do not expect the instructor to tell you the "correct answers." The instructor will verify your final conclusions and check over your analysis to determine if you chose appropriate supporting evidence. The important point to remember is that you wish to make certain conclusions regarding the nature of your unknown and to support these conclusions with a logical framework of evidence.

2. SOME EXPERIMENTS

One of the interesting problems of plant transport concerns the mechanism of water movement. Three processes might be important: transpiration, root pressure, and capillarity. One experiment which you might attempt, then, is to evaluate these three processes as possible mechanisms for moving water through the xylem. (You might wish to demonstrate that water moves through the xylem as opposed to the phloem.) If you decide to conduct such an experiment, you should first read your text and come to a clear understanding of what transpiration, root pressure, and capillarity are. Then outline an experimental procedure including a hypothesis, materials, and methods, and discuss your plans with your laboratory instructor. The instructor will first determine if the experiment can be performed with the time and materials available. The instructor will also discuss with you any possible errors of design. It is not the instructor's responsibility to tell you what to do, but to guide you in the right direction. For example, he may suggest that you do not have an adequate control. He will not tell what control to establish; it is your task to reexamine your procedure and discover where a control is needed. If you become stymied, approach your instructor again. By means of a dialogue, the instructor will try to lead you to the correct solution of your problem.

There are numerous other problems you might investigate. What sorts of environmental factors affect the rate of transpiration? Does transpiration occur at the same rate in different kinds of plants? Does the rate of transpiration vary if the roots of a series of plants are put into different salt solutions? Does temperature or light affect root pressure? Is root pressure greater in young plants than in old plants?

3. REPORT

Prepare a written report of your work. Use the format described in Exercise 3.

SUGGESTED READINGS

Jensen, W. A. and Kavaljian, L. G. (1966). "Plant Biology Today," 208 pp. Wadsworth, Belmont, California. Paperback.

Steward, F. C. (1964). "Plants at Work," 184 pp. Addison-Wesley, Reading, Massachusetts. Paperback.

EXERCISE

10 FUNCTIONAL ORGANIZATION OF ANIMALS

Like that of plants, animal organization must be able to solve the three essential problems of maintenance: obtaining energy, exchange with the environment, and internal transport. If a plant may be thought of as a device for converting light energy to chemical energy, an animal may be thought of as a device for obtaining chemical energy from its environment. In other words, plants and animals differ fundamentally in their organizational bases for obtaining energy. There is great structural and functional diversity among animals, but most of this diversity may be generalized into broad patterns.

Most free-living animals obtain chemical energy in the form of organic molecules which we call food. Food may be gathered by means of one of three general patterns.

First, animals may actively search for food. Food seekers usually have well-developed sense organs, locomotory organs, and nervous systems. These animals may become food specialists. The crayfish and the frog, which will be studied as examples of animal organization, are examples of food seekers.

Second, an animal may pass the external medium through the alimentary canal. The substrate swallowers live in soil, mud, or muddy sand. The earthworm is an example of a substrate swallower.

Third, an animal may be a filter feeder. Most filter feeders set up a current of water that passes over some part of the body which traps the food on the basis of particle size. Particles too large are rejected; particles too small are not caught. Most filter feeders are sessile.

Because the energy intake of animals is in the form of organic molecules, they require a digestive mechanism. Usually this mechanism is expressed morphologically as an alimentary canal. Although the detailed morphology of the gut varies, the functional divisions are similar in all animals. The following five functional regions are readily recognized:

1. Mouth and region immediately posterior to it
    a. Region of food intake

   b. Often assisted by appendages such as tentacles and internal structures such as teeth
   c. May have a muscular pharynx
   d. Some digestion may start here
2. Esophagus
   a. A tube for passage of food to the stomach
   b. May enlarge at posterior end into a crop for storage
3. Stomach
   a. Region of further physical breakup of food. For this action part of the stomach may form a muscular gizzard or may contain an internal tooth mechanism
   b. Digestion begins here in most animals
4. Intestine
   a. The region of most of the digestion and absorption
   b. May be associated with diverticula known as digestive glands or liver. Digestion may occur in these glands or they may secrete digestive enzymes into the intestine
5. Rectum
   a. The absorption of water and the formation of feces occurs here

Transport mechanisms are necessary in animals for the same basic reason that they are necessary in plants. An increase in size and/or activity requires a more rapid distribution of materials than diffusion alone can satisfy. Thus there are mechanisms that bring materials within diffusion distance of the cells where the materials are used. In animals the basic means of achieving transport is by circulating an internal fluid. The fluid may be enclosed in tubes or it may occur in the space between the gut and the body wall. The fluid may be propelled by a pump, commonly called a heart, or the fluid may be moved about by cilia or by muscles of the body wall.

Exchange is necessary in part because most animals respire aerobically, thus removing oxygen from the environment and releasing carbon dioxide to the environment. These gases diffuse readily across differentially permeable membranes. Membranes permeable to gases are also permeable to water. Thus animals, like plants, have a problem of water balance. Salts also may move across an exchange surface. Exchange in animals may be one of the functions of the outer body layer or integument. Many animals have an organ specialized for exchange, such as the gill, the lung, or the tracheal system.

Only part of the metabolic by-products are exchanged across an exchange surface. Most animals have an additional excretory system whose main organ is the nephron.

Animal organization will be examined in some detail in the crayfish and frog. These two animals represent two great major subdivisions of animals, invertebrate and vertebrate. If you have previously dissected either of these animals, do not repeat that work here, but continue immediately with your study of the unknown animal.

1. THE CRAYFISH (Figs. 10-1, 10-2)

The crayfish is an invertebrate; that is, it lacks an internal skeleton composed of vertebrae. Invertebrate is not a fully satisfactory term, for it denotes the absence of a structure, it applies to most animals, and we might expect greater uniformity than actually exists. The crayfish is not the most specialized nor the most primitive of invertebrates but is a member of an advanced group of animals, the phylum Arthropoda. This phylum includes the lobster, shrimps, insects, spiders, crabs, millipedes, centipedes, etc. All of them have jointed appendages. Thus the crayfish represents an animal fairly high up on the evolutionary scale, but one which is not highly specialized for a particular mode of life. Of course one should keep in mind that all animals are specialized to at least some degree. As you study your crayfish, record your observations in the spaces provided. Record anything you observe, not just those things requested in the laboratory directions.

A. Observations of a Living Crayfish

NOTE ON DIRECTIONS: When dissecting an animal, one needs to know where to look to find a structure. Basic directions are related to the major axis of the animal body, the longitudinal axis. The *anterior* end is the head end; *posterior* refers to the tail end. The top or back side is *dorsal;* the belly or bottom is *ventral. Medial* refers to the midline of the body; *lateral* refers to the sides. *Proximal* parts are nearer the main axis; *distal* parts are farther from the main axis.

Examine a living crayfish in a bowl of water. At the anterior end, you can see a structure below the head which is beating up and down rapidly. This structure, which is part of an appendage of the mouth, pulls a current of water over the gills which are located under the chitinous exoskeleton above the legs. The current of water carries oxygen to the gills and removes carbon dioxide. Note that the crayfish consists of two general regions, an anterior unsegmented region and a posterior segmented *abdomen.*

Place a small piece of *Elodea* or pellets of commercial fish food in the dish with the crayfish. Watch the crayfish for a few minutes to see if it will eat. If it does, observe and record the operation of the mouth parts.

The mouth parts are modified appendages. The crayfish has 19 pairs of appendages: 1 pair of antennae, 1 pair of antennules, 6 pairs of mouth parts, 5 pairs of walking legs, 5 pairs of abdominal appendages, and, at the posterior, 1 pair of uropods. The antennules and antennae are sensory structures. How does the crayfish hold these?

The appendages will be examined in more detail later. Are the large chelae (or pincers) used to manipulate the food? If not, what might be their function? If you poke a pencil at a crayfish, you may find the answer.

B. Transport: The Heart

Place a crayfish in a dissecting tray and remove the chela and walking legs by cutting them off close to the body. Save at least one chela and one walking leg for later study.

Carefully remove the top half of the covering over the thorax, that part of the body to which the legs were attached, and expose the *heart*. The crayfish has a partially open circulatory system. The blood leaves the heart in *vessels* and returns through spaces of the body cavity. A large space or *sinus* surrounds the heart. Blood enters the heart through openings called *ostia*. When the heart contracts, valves close the ostia to prevent backflow. If you remove the covering carefully, you will be able to observe the action of the heart.

C. Exchange: Gills and Nephron

Next examine the gills. Remove the covering from one side of the body to expose the gills. Although the gills are protected to a degree, they are exposed to drying and are located on the surface where water can pass over them. How are the gills arranged in the gill chamber? Does the arrangement facilitate gaseous exchange between the gills and water passing over the gills?

At the base of the antenna (external) a small pore can be seen. This is the opening of the green gland, the excretory organ or nephron of the crayfish. Internally the organ lies just posterior to the stalked eyes. Carefully remove the carapace posterior to the eyes and expose the excretory organ. By careful dissection you should be able to find a ventral, glandular part connected posterodorsally with a more dorsal bladder. The glandular part is firm and shaped like a convex lens. The bladder part is easily overlooked as it is thin and easily collapsed.

D. Reproductive Organs

In the crayfish, the reproductive organs or *gonads* lie in the body cavity near the heart. The male organs or *testes* usually lie just below the heart and appear as two white lobes. A sperm duct conducts sperm from the testes to the base

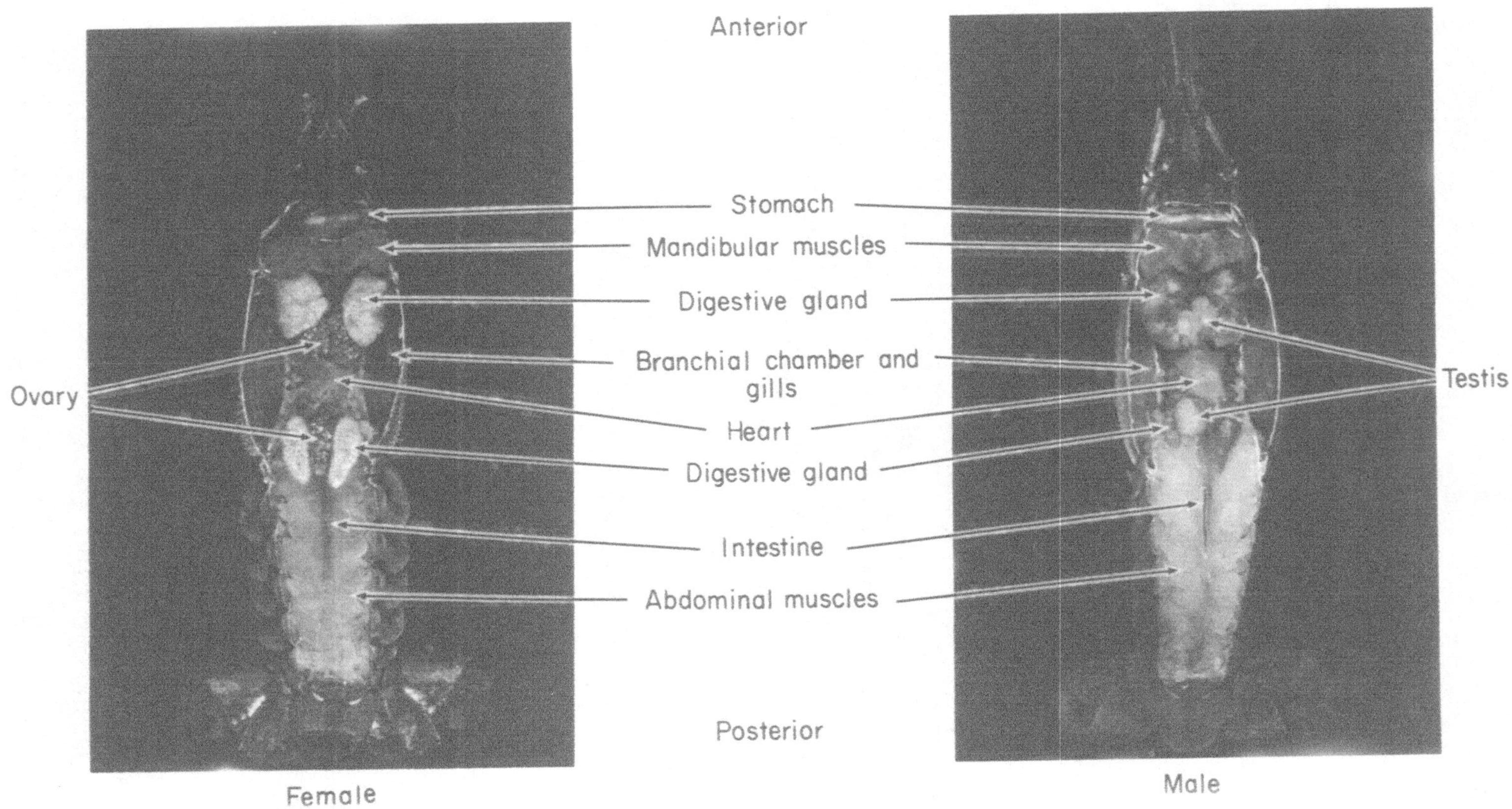

FIG. 10-1. (Left) Internal anatomy of a female crayfish. (Right) Internal anatomy of a male crayfish.

of the last walking leg. The *ovaries*, or female gonads, are usually ribbonlike and may be much more extensive in size, depending on the time of year. Oviducts conduct eggs to the base of the third walking leg.

E. Digestive System

As you examine the digestive system, relate the structures to the functional regions described in the introduction. The mouth region is ventral and contains mandibles for crushing particles of food. A short *esophagus* leads to the *stomach*. Further grinding of food occurs in the stomach. Cut it open and locate the hardened *gastric mill*. Dissect carefully and you can determine how the gastric mill operates. Most of the digestion occurs in paired *digestive glands*. These yellowish, greenish or brownish structures occupy most of the anterior part of the body cavity on either side of the gut. The digestive glands also serve as a food storage depot. Undigested *residue* passes through the narrow *intestine* and is eliminated through the anus.

The intestine lies along the dorsal side of the well-developed muscles of the abdomen. Note that the abdomen is mainly muscle, the source of gastronomical delight in shrimp and lobster.

F. Nervous System

The nervous system is ventral in the invertebrates. Carefully remove all the body organs except the esophagus. The ventral *nerve cord* will be seen on the midventral line. The nerve cord of the abdomen consists of six pairs of ganglia with their branches and connections. Note that the posterior pair of ganglia which innervate the uropods are enlarged. The addition of methylene blue to the nerve cords will make them easier to see.

The brain or *cerebral ganglia* are found on the dorsal surface above the anterior to the esophagus. These ganglia innervate the eyes, antennae, and antennules. From the cerebral ganglia, two nerve cords pass around the esophagus and connect with the anterior ganglia of the ventral cord. The anterior part of the nerve cord may be difficult to see because ridges of the body wall conceal it. However, the nerve cord is easily seen underneath the abdominal muscles. The great number of ganglia increases the complexity of nervous organization. This increase in complexity of nervous organization has evolved along with the increased complexity of responses. The increased complexity of responses is evident, in part, in the nature of the appendages. The jointed appendages make possible a high degree of manipulation and control of movement.

G. The Appendages

Examine the appendages (see also Fig. 10-2). The *walking legs* are five-jointed. The first walking leg has the distal two segments modified into a

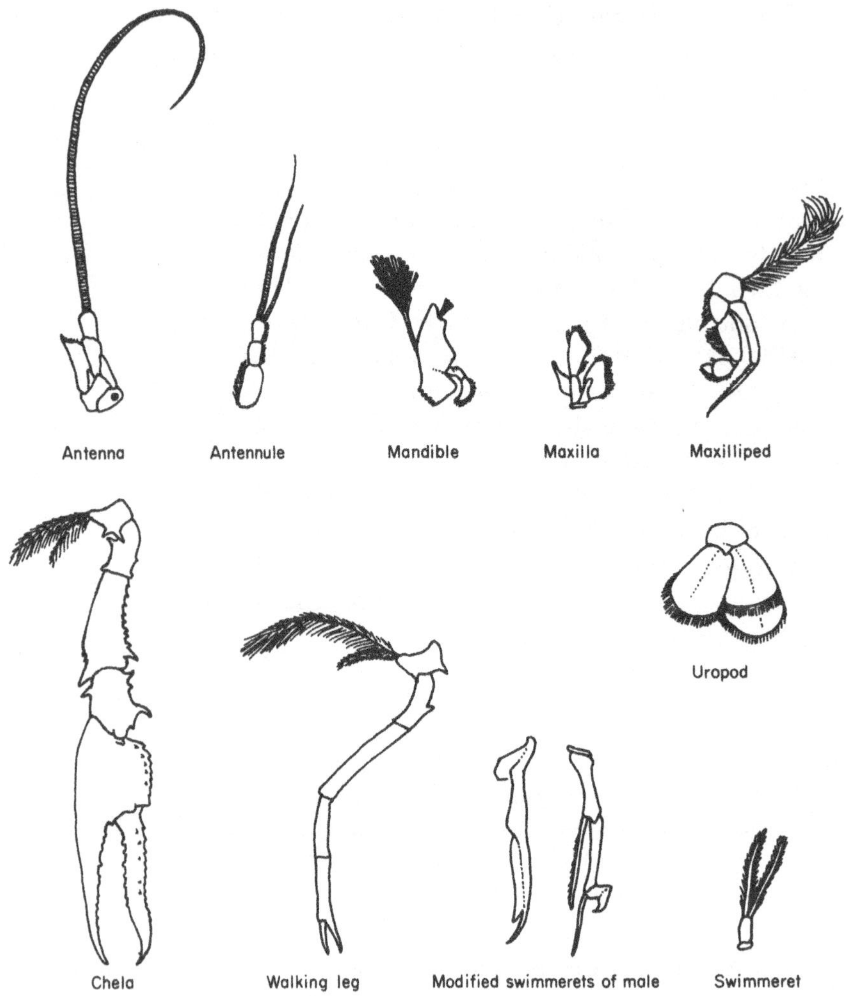

FIG. 10-2.
The appendages of a crayfish.

chela. Orient the chela so that the movable segment is on your right. Carefully remove the exoskeleton from the broad area of the chela proximal to the two "fingers" and expose the underlying muscle. One small muscle runs from the movable appendage along the lateral edge (the right side if you orient it as directed above). A large muscle runs from the movable segment into the broad base where it fills most of the chela. Carefully grasp the small muscle with your forceps and gently pull it. In which direction does the appendage move?

Repeat with the large muscle. Which muscle is likely to be capable of exerting the greater force?

Does this correlate well with the use to which crayfish puts the chela?

What does this system indicate about the theory of opposed muscles?

All the remaining appendages are biramous; that is, each has two branches. The appendage consists of a basal pair of segments which in turn bears two branches at its distal end. These appendages and their parts are variously modified for particular functions. Note that the sensory *antennae* have one branch greatly elongated and the sensory *antennules* have both branches elongated. Of what advantage to the animal is the elongation of these structures?

Carefully remove the mouth parts. Starting at the outer surface, the first three are *maxillipeds*. The maxillipeds are walking legs modified for the manipulation of food. Examine these appendages carefully for structural modifications that fit their function. Beneath the maxillipeds are two pairs of *maxillae* that are also modified for food handling. Are they structured similarly to the maxillipeds? The internal mouth part is the *mandible*. How is it specialized for breaking up food particles?

How do the *uropods* differ from the other appendages? The uropods and an outgrowth of the last abdominal segment, the *telson*, form a tail fan. How is the tail fan used? Ask the laboratory instructor to demonstrate the startle reaction. How does this behavior correlate with the muscular development of the abdomen and the enlarged posterior ganglia? Also examine the structure of the abdominal limbs.

The abdominal limbs have two functions. The first two are modified in the male for copulation. In the female, the eggs are carried on these appendages.

H. Chromatophores

Chromatophores are effectors that contain pigment that is capable of changing its position inside the chromatophore.

Two finger bowls are painted, one black and one yellow. Two small crayfish are placed in each bowl and illuminated for several hours. Examine one crayfish from each bowl under a binocular microscope. Determine the degree of dispersion or concentration of the pigment in each of the kinds of chromatophores. The chromatophores can be seen best on the uropods. The pigment may form a small rounded spot (most concentrated), a widely spreading and branching network (most dispersed), or some intermediate condition. On the basis of your observations, can you come to any conclusions concerning the function of the chromatophore system of this crayfish?

2. THE FROG (Fig. 10-3)

All vertebrates share certain characteristics. The vertebra, a small bone that surrounds the spinal cord, is the common feature from which the name vertebrate was derived. The Vertebrata is a subphylum of animals which includes the fish, frogs, toads, salamanders, snakes, lizards, turtles, birds, and mammals, including man. We will examine the frog because it is readily available, inexpensive, and illustrates the general body plan of vertebrates.

Before you begin your examination of the frog, remember the emphasis of your study is not learning names of structures for the sake of learning names, but to understand how a frog is structured to form a living system. If the frog is to survive as a living entity, it must carry out the same processes of maintenance as the crayfish. Keep in mind that although the frog has excretory, digestive, and transport systems, these systems may be based on functionally similar but structurally different organs. Compare such organs as the heart, stomach, and excretory organ, for example.

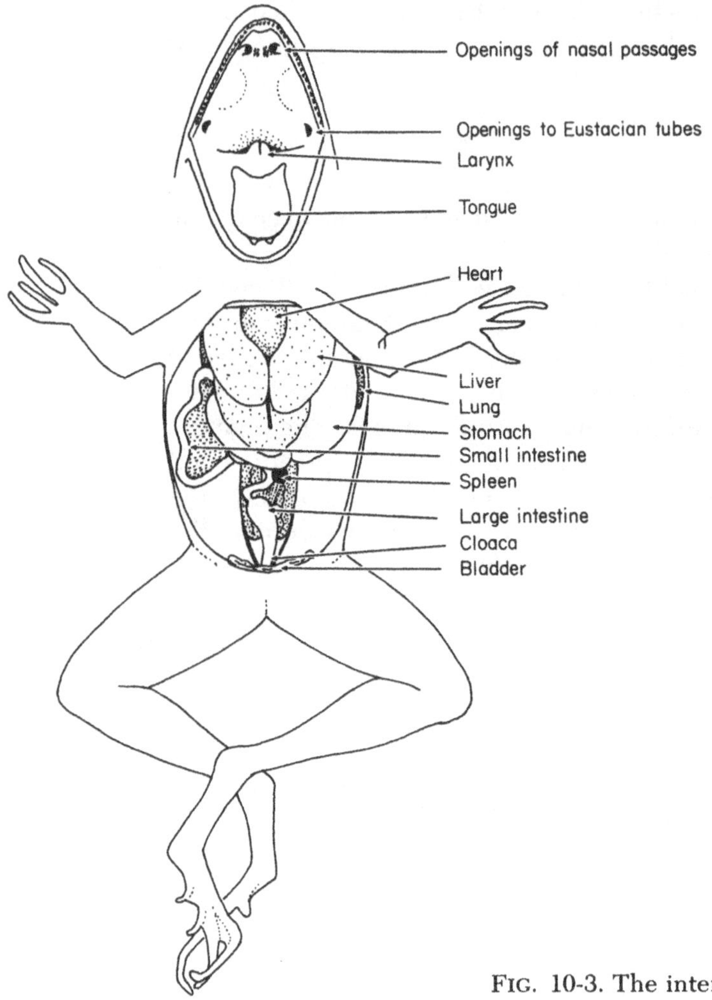

FIG. 10-3. The internal anatomy of a frog. Ventral view.

A. Observations of a Living Frog

Examine the living frog in the water. Note the location of the *eyes*, the *mouth*, the *external nares* (nostrils), and the *tympanic membranes* (eardrums). The frog has no projecting external ear as we do. Where do the eyes, tympanic membranes and external nares lie with respect to the water level? Can you see any movements of the mouth? What is the significance of the position the frog assumes in the water and of the movements of the mouth? The eyes are covered by thick *eyelids*. Inside the lower eyelid is a thin *nictitating membrane*, an extension of the lower eyelid, which is pulled up over the eye when the frog submerges, permitting some vision and affording some protection to the delicate surface of the eye. Cause the frog to submerge and observe the functioning of the nictitating membrane.

B. External Morphology and Mouth Cavity

Before you begin the dissection of your anesthetized or preserved frog, examine any of the structures mentioned above that you could not readily observe. Also examine the structures of the mouth. Examine the *tongue*. Where is it attached? What is the shape of its tip? Feel along the edges of the jaws for *teeth*. Do you find them in both jaws? There are also two sets of teeth located well forward in the roof of the mouth. What might their function be?

The openings from the nasal passages into the mouth lie lateral to the two sets of teeth. The back of the mouth is called the pharynx. A little mound, the *larynx*, is located in the midline of the pharynx. In the larynx is a longitudinal slit, the *glottis*, through which air passes on its way to the lungs. The *esophagus* opens just behind the larynx. Laterally, just a little above and behind the angle of the jaw, are found the openings to the *Eustachian tubes*. Each tube leads upward into the cavity of the middle ear just beneath the tympanic membrane.

Would you expect a frog to swallow food and air at the same time? Do you?

C. Transport: The Heart

The transport system is more complex in the frog, especially in the development of blood vessels. These will not be described, but careful dissection will reveal many of them. They may be readily traced from the heart.

Place your frog dorsal side down in the dissecting tray. Carefully open the body cavity or *coelom* by making a longitudinal cut just lateral to the midline and extending from the posterior to the anterior legs. Keep the points of the scissors shallow so that the organs underneath are not damaged. At the posterior end of the longitudinal slit locate the ventral abdominal vein. Carefully cut a tongue of skin to which the vein is attached and fold the tongue of skin and vein posteriorly. Then at each end of the longitudinal slit make lateral cuts toward each leg so that the skin and muscle layer of the body wall will open up like doors. Pin these flaps back and proceed with the examination of the internal organs.

Find the *heart* in the center of the breast region. It will probably be necessary to cut through the bony structures ventral to the heart in order to expose it properly. How does the beat of the heart differ from that of the crayfish? The large lobes of the *liver* more or less surround the heart. If you have a female frog, the body cavity may be filled with ovaries containing black and white eggs. If this is the case in your specimen, carefully loosen the egg-bearing tissue of one side from the other organs and remove it from the body.

D. Digestive System

The frog and the crayfish have several similarities in the digestive system as well as some conspicious differences. With the aid of Fig. 10-3, locate the structures of the digestive system and determine their role in the functional organization of the system.

Projecting from under the liver on the frog's left side is the *stomach*. Anteriorly the stomach is joined by a tube, the *esophagus*, which connects the pharynx with the stomach. At the posterior end of the stomach there is a narrow constriction where the *small intestine* connects to the stomach. The small intestine coils considerably and empties into the *large intestine*. The large intestine is a large, dark-colored tube that runs dorsally and posteriorly toward the anus. The posterior section of the gut is the *cloaca*, a tube that receives the large intestine and ducts from the gonads and kidneys. Does the stomach have a gastric mill? Does the crayfish or frog have a more complex intestinal region?

The arrangement and number of digestive glands is more complex in the frog. The large three-lobed liver secretes bile which may be stored in the gall bladder, a small greenish sac located between the middle and right lobes of the liver. Another digestive gland is the *pancreas*. This gland is located in the *mesentery* (the sheets of connective tissue that support the organs) between the stomach and the small intestine. Mesentery was also present in the crayfish, though not specifically mentioned. The pancreas is a source of enzymes, such as amylase. Note that digestion occurs primarily in the stomach and intestine in the frog and not in a digestive gland as in the crayfish.

Near the anterior end of the large intestine is the *spleen*, a spherical, reddish structure attached to the mesentery. In the frog, the spleen is a blood-forming organ. No comparable organ is known in the crayfish. The spleen properly belongs to the transport system.

E. Exchange: Kidney and Lung (Fig. 10-4)

Along the dorsal body wall lies a space, the subvertebral lymph sinus. Lying in this sinus on either side of the posterior portion of the backbone are the reddish-brown *kidneys*. Along the middle of the ventral surface of each kidney is a lighter yellow or orange streak, the *adrenal gland*, which secretes hormones. Along the outer edge of the kidney is a duct which carries urine from the kidney to the cloaca. On the ventral surface of the cloaca, just anterior to the *anus*, a thin-walled sac, the *urinary bladder*, is attached. This sac is usually collapsed and not conspicuous. In both the crayfish and frog the bladder stores fluid. But note that the structural relationship between bladder and kidney (or excretory organ) differs in the two animals as does the pathway for the elimination of urine.

At the anterior end of the kidney is a patch of yellowish/orangish tissue with fingerlike lobes, the *fat body*.

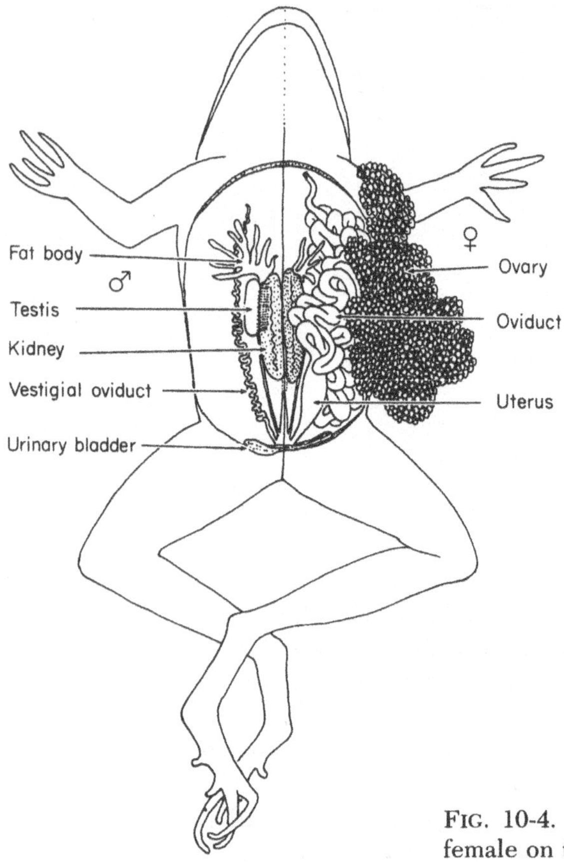

FIG. 10-4. The urogenital system of a frog. Male on the left female on the right.

Lateral to the esophagus in the extreme anterior end are the *lungs*. Can you think of why lungs are advantageous to a terrestrial animal?

Frogs are frequently heavily parasitized. Open the lungs or intestine and examine these tissues for parasites. The most likely parasites will be flatworms, related to the free-living planarians. The parasite may be mounted on a slide and examined under the microscope. You may wish to sketch its appearance.

F. Reproductive System (Fig. 10-4)

The reproductive systems of the frog follow the same basic plan as those of the crayfish, except they are more complex.

If the frog is a male, the *testes*, whitish, elongate bodies about the size of a small bean, will be located ventrally at the anterior ends of the kidneys. Lateral to the kidney is a small wavy tube running posteriorly to the cloaca. This tube is a *vestigial oviduct*, which has no function in the male, but corresponds to the functional oviduct of the female.

If the frog is a female, the *ovaries* are likely to be distended with partly developed black and white eggs. The eggs are released into the body cavity where ciliated tracts sweep them into the *oviducts*. The openings of the oviducts are beside the esophagus near the front end of each lung. The oviducts are long and coiled; the walls secrete jelly, which is applied to the eggs as they are moved through the tube. At its posterior end, the oviduct connects with the thin-walled *uterus*, which stores eggs until a large number have accumulated and are ready to be laid. The posterior end of the uterus connects with the cloaca.

G. Demonstration of Skeleton

Examine the demonstrations of the skeleton of the frog. Note how the skeleton is organized to give support. Contrast the supporting systems of frog and crayfish. Does each system contribute to the functional organization of the whole? Which system is better? Is "better" a meaningful word in this context?

3. THE NERVOUS SYSTEM (OPTIONAL)

If time permits, dissect out as much of the nervous system as possible. There are 10 pairs of *spinal nerves* in the body cavity. These can be observed by removing all the viscera from the frog. The nerves, for the most part, will appear as silvery bands emerging from the spinal cord and passing along the dorsal body wall. The second spinal nerve is large and enters the foreleg or arm. The first spinal nerve is smaller and slightly anterior to the second. The third is also smaller and slightly posterior. Branches of the second and third fuse to form the *brachial plexus.* A *plexus* is the place where nerves connect and exchange fibers. The fourth, fifth, and sixth nerves are about the same size as the first and third. They run diagonally on the dorsal body wall in a posterior direction. The seventh, eighth, and ninth nerves are much larger and run in an almost direct posterior direction. These three nerves exchange fibers to form the *sciatic plexus.* The tenth nerve emerges from the urostyle (tail bone). This nerve is the smallest and lies parallel and medial to the ninth. It may also contribute to the sciatic plexus.

Two thin *autonomic trunks* lie on either side of the spinal column (backbone) on the dorsal body wall. They have small threadlike branches, the *rami communicantes,* that connect them to the spinal nerves. Under the low power magnification of the binocular dissecting microscope, it is possible to observe the *trunk ganglion* at the place where a ramus communicans connects to the autonomic trunk.

The *brain* may be exposed by carefully cutting away the bone overlying the brain. First remove the skin and muscle from the dorsal region of the head and trunk. Shave off a little of the top of the skull between the eardrums and carefully make a hole. Work forward from this point. Insert the scalpel point under the free edge of the opening and chip outward. If proper care is exercised, small pieces of bone may be removed without damaging the brain. Chip away the entire roof of the cranium as far anteriorly as the anterior end of the *olfactory nerves*. These nerves attach to the most anterior part of the forebrain and run forward into the nasal capsules.

Chip posteriorly to the first vertebra. Insert the sharp tip of your scissors into the spinal canal between the spinal cord (nervous tissue) and the spinal column (bone). Cut along each lateral wall of the spinal column so that the roof may be removed to expose the spinal cord in place.

Just posterior to the olfactory nerves are the paired *olfactory bulbs* and paired elongate, slender *cerebral hemispheres*. These structures lie between the eyes. Posterior to the cerebral hemispheres are two enlarged structures, the *optic lobes*, which form the dorsal part of the midbrain and lie between the tympanic membranes. The rest of the brain consists of a narrow transverse flap, the *cerebellum*, and the *medulla oblongata*, the latter covered over by a dark layer of tissue, the *posterior choroid plexus*.

You may now remove the entire central nervous system. Be especially careful to tease away the ventral attachments of the brain from the underlying bone. As you lift the brain anteriorly, it will be necessary to cut the olfactory tracts at their distal ends. As you lift up the cerebral hemispheres, you will see the optic tracts attached to the brain. Posterior to these tracts there is a small white body, the *pituitary gland*. Special care must be taken to remove the pituitary gland intact. After the brain is removed, place it in water.

On the dorsal side, behind the cerebral hemispheres lies the *diencephalon*. It occupies a relatively small area. Its dorsal surface is covered by the *anterior choroid plexus*. A dorsal outgrowth of this region is the *pineal body*.

On the ventral side the optic tracts cross as they enter the brain to form the *optic chiasma*. Behind the chiasma is the floor of the diencephalon, the *hypothalamus*, an enlarged bilobed structure. The pituitary gland is attached posteriorly to the hypothalamus.

4. CELL TYPES AND TISSUES OF ANIMALS

The functional organization of animals requires only a few basic cell types. Although there may be extensive patterns of variation of ultrastructure (the structure determined by the electron microscope), the patterns seem to be organized into only a few cell types or tissues. Cell types are essentially supportive, protective, conductive, contractive, and secretive.

A. Supportive (Connective)

Supportive or connective tissue is characterized by the presence of intercellular material secreted by the cells. Note how the structure of each of the four kinds of connective tissue is related to its function. There are other modifications of the connective tissue plan, but we will not examine them.

1. *Fibrous.* This tissue is characterized by the presence of many fibers. It essentially is a binding tissue which loosely holds together other tissues to form organs and supports organs in the body cavity (see Fig. 10-5).

2. *Cartilage.* This tissue is one of the two main supportive tissues. The cartilage cells secrete a hard, rubbery matrix and are grouped in two's or four's in small cavities in the homogeneous matrix (see Fig. 10-6).

3. *Bone.* This tissue is the other main supportive tissue. The matrix consists of organic material and calcium salts. Bone is not a solid structure, but contains a system of channels known as Haversian canals. Blood vessels and nerves which supply and control the bone cells course through the canals. The matrix is secreted in concentric rings around the canals and the bone cells are in cavities in the concentric rings. Small canals connect the bone cells to each other and to the Haversian canals (see Fig. 10-7).

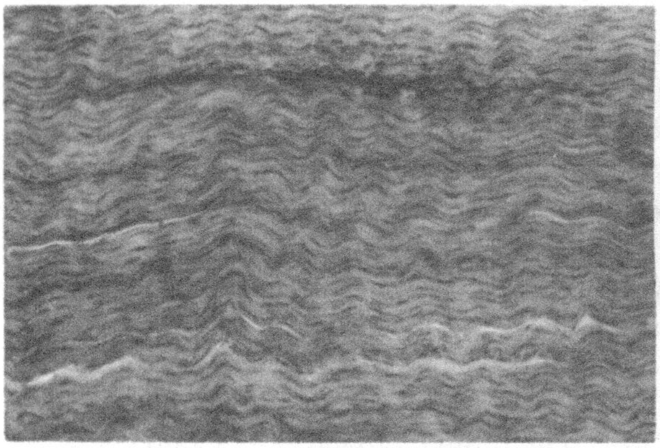

FIG. 10-5. Fibrous connective tissue.

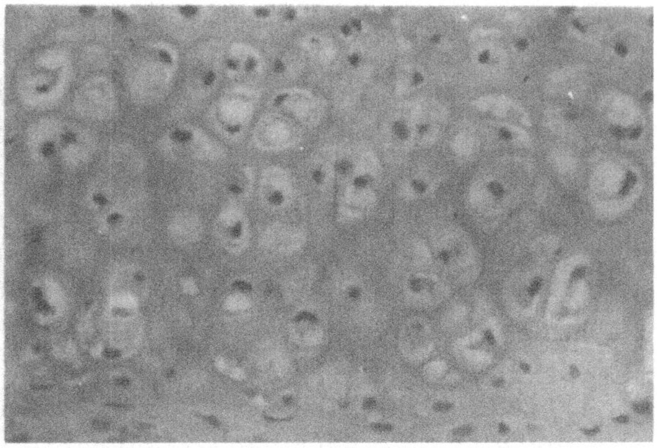

FIG. 10-6. Cartilage.

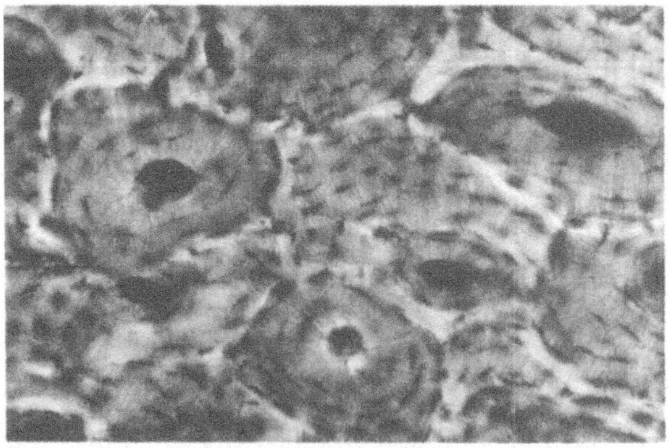

FIG. 10-7. Bone.

4. *Blood* is a connective tissue in which the intercellular material is fluid.

Two major types of cells are present, the red blood cells which contain hemoglobin and function in the transport of oxygen and carbon dioxide and white blood cells which function to destroy foreign particles in the blood and body fluids and perhaps to form antibodies. Human red cells are nonnucleated and are smaller than the white cells. Red cells are about 7–8 microns in diameter and white cells range from 8–20 microns in diameter.

B. Contractive (Muscle)

If any tissue is characteristic of multicellular animals, it is muscle. The significance of muscle lies in its capability of shortening and exerting a pull. The muscle must contract against something, otherwise no pressure could be exerted. Thus muscle is associated with skeletal units to form a muscle-skeletal system. You examined such a system in the crayfish chela. Examine slides which demonstrate the two basic types of muscle.

Skeletal muscle has striations and is found where rapid, strong movement occurs (see Fig. 10-8). *Smooth muscle* lacks striations and is found where slow sustained movements occur (see Fig. 10-9). What other differences can you find between these two kinds of muscle cells? You may be interested in learning more about the cellular nature of muscular contraction. We still have much to learn, even though much progress has been made in recent years, especially in striated muscle.*

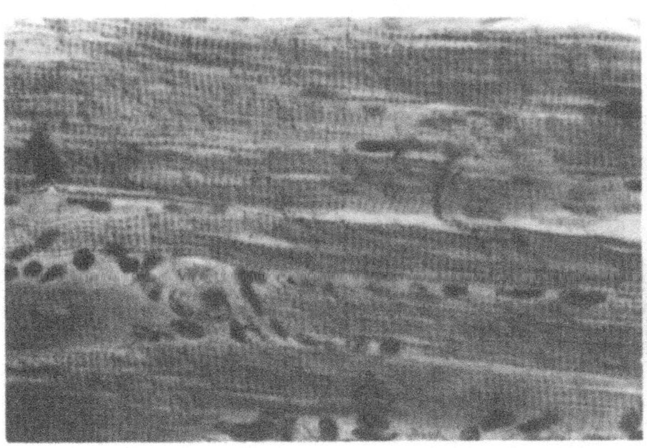

Fig. 10-8. Striated muscle.

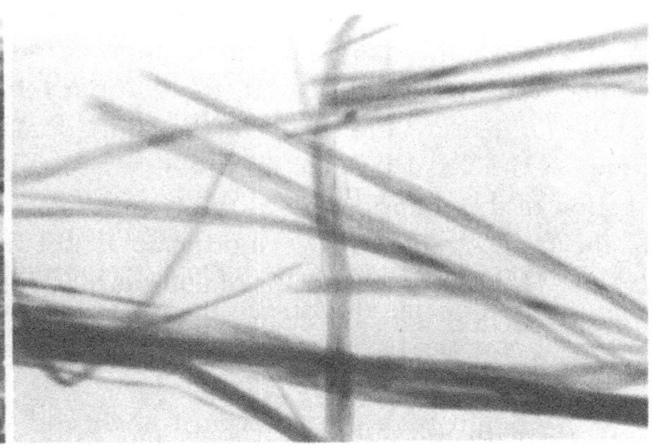

Fig. 10-9. Smooth muscle. The fibers have been teased apart.

* See H. E. Huxley, The contraction of muscle, *Sci. Am.* **199** (November), 66–72 (1958).

C. Conductive (Neuron)

The other tissue which is especially characteristic of multicellular animals is *nervous* tissue.

This tissue receives a stimulus and transmits a message. The basic cell is the *neuron*. A neuron consists of a cell body, which contains the nucleus, and two or more *neurites* of "fiberlike" processes extending from the cell body. You have seen concentrations of neurites which are called nerves. Concentrations of the cell bodies of neurons are called ganglia. Where did you see ganglia?

5. THE ORGANIZATION OF AN ORGAN

We have examined animal organization at two levels, the level of the whole animal and, to a small extent, the cellular level. The organs which form the whole animal are the result of the functional integration of the various cell types or tissues that make up the organ. We will examine a slide of the cross section of the pyloric region (the region nearest the intestine) of the stomach of the frog. Again, what is significant is not the names of the various tissue layers. The names are means of communicating information. But the information associated with the name should primarily consist of the understanding of how the tissue layer is organized and what its contribution to the integrated function of the whole organ is.

Begin with the outermost layer and work your way inward as if you were taking a trip through a strange country. Let Fig. 10-10 be your map.

The *peritoneum* is a thin layer of flattened cells. There are occasional broader areas where the nuclei of the cells are located. Below this layer are a few connective tissue cells and fibers that are not easily distinguished in the section. The peritoneum is an epithelial tissue. Where did you previously see epithelial cells? What might the occurrence of epithelial cells suggest about their function?

The *longitudinal layer of smooth muscle* may be poorly developed and thin. Because the cross section is cut at right angles to the length of the fibers, only their cut ends will show as small roundish or slightly irregular spots.

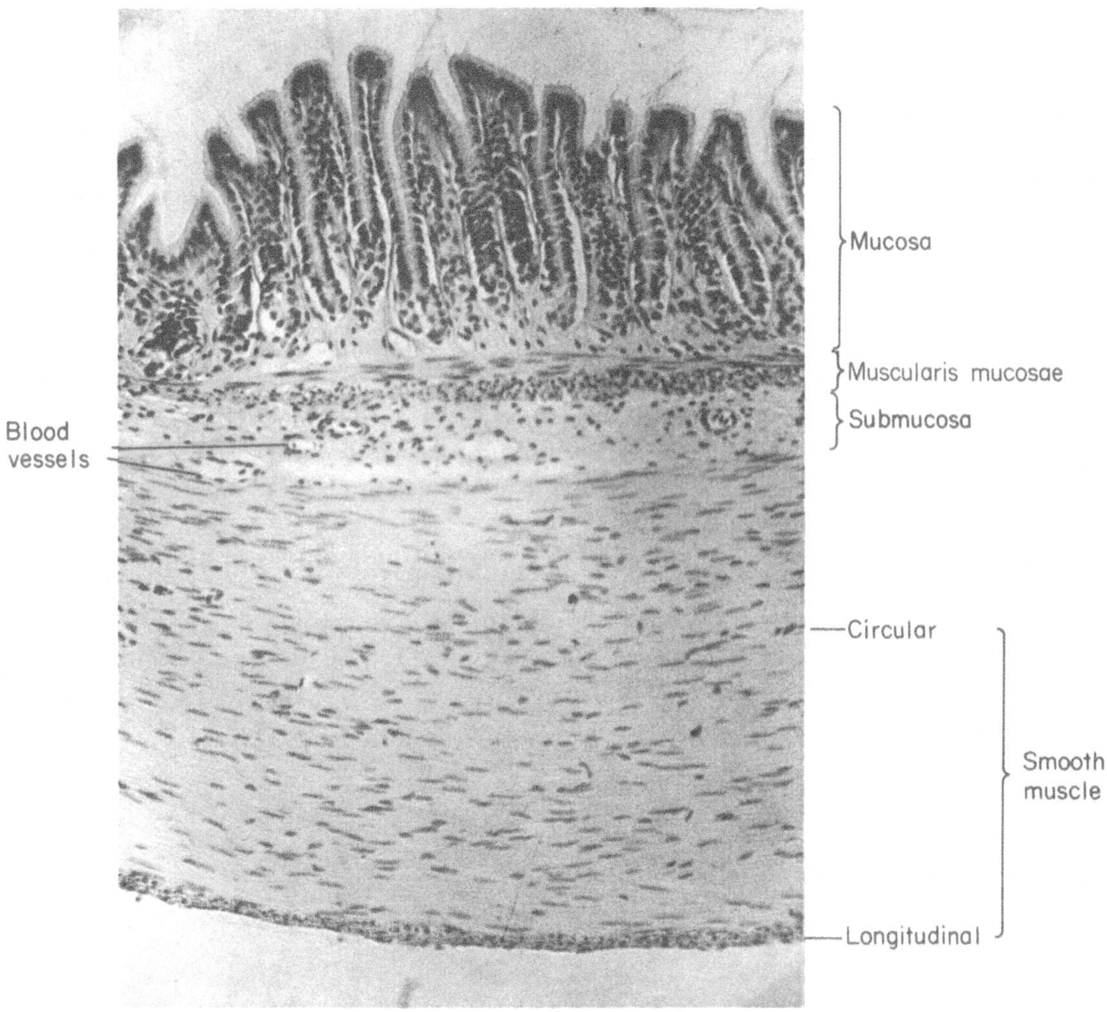

FIG. 10-10. Cross section through the stomach of a frog.

The *circular layer of smooth muscle* is thick to give strength for the internal physical breakup of food. Because the cross section is cut parallel with the fibers, whole cells lying parallel and overlapping each other and with a nucleus near the middle of each cell, can be seen. Also, sections of fine *capillaries* are scattered through this layer.

Using your knowledge of muscular contraction, work out the functional system by means of which these two muscular layers move food through the stomach. The same layers are present also in the esophagus and intestine.

The *submucosa* forms a strengthening layer and supports *blood vessels* that pass through to the inner layers. What kind of tissue occurs in this layer?

The *muscularis mucosae,* consisting of both longitudinal and circular layers of muscle, moves the soft inner tissue.

The *mucosa* is the internal, secreting layer of the stomach. It is a complicated layer, consisting primarily of a *glandular epithelium,* and connective tissue which binds the cells together. This arrangement of tissues is a good illustration of the way in which connective tissue penetrates other tissues as a binding or enclosing agency to hold cells in their proper relations to each other.

How many of the following cellular types are present in the cross section? Secretive, conductive (present, though probably not detected), contractive, connective or supportive, protective.

SUGGESTED READINGS

Griffin, D. R. (1962). "Animal Structure and Function," 120 pp. Holt, New York. Paperback.

Russell-Hunter, W. D. (1968). "A Biology of Lower Invertebrates," 181 pp. Macmillan, New York. Paperback.

Russell-Hunter, W. D. (1969). "A Biology of Higher Invertebrates," 224 pp. Macmillan, New York. Paperback.

EXERCISE

11 ANIMAL UNKNOWN

In the previous exercise, the principles of animal organization were introduced and illustrated by dissecting two animals, an invertebrate, the crayfish, and a vertebrate, the frog. The purpose of this exercise is to permit the student to investigate some aspects of the structure or function of animals which have not previously been studied.

I. MORPHOLOGICAL

A large variety of animals representing the major groups of the animal kingdom are available. If you have an interest in some particular group of animals, you may request a member of that group for your study. If you have no special interests, the instructor will assign you a specimen. The general features of the animal should be determined on the basis of what you learned from Exercise 10. However, there may be special structures with which you have had no previous experience. You should attempt to learn the function(s) of such structures by observing living organisms. If necessary, make up names for such structures based on their appearance, structure, and function.

Prepare a report following a logical format of your own choosing in which you describe your animal. The report may include drawings of the morphology and any description of the activity of the living animal. Because you are deducing the names and functions of structure based on their similarities to structures previously learned, the report should include your reasons for assigning names and functions.

After your report is read by your laboratory instructor, it will be returned to you with references so that you may then compare your work with the accepted description of the animal.

2. FUNCTIONAL

There is, of course, an immense variety of functional problems which might be investigated. Many of these will require more time and equipment than normally is available in a general biology course. Some suggestions of possible investigations follow.

A. Respiration

In Exercise 6 some whole animal studies of respiration were suggested. If you did not pursue any of those studies or if you wish to explore some of the problems in greater depth or with different organisms, the investigation of oxygen consumption is an appropriate topic for this exercise. Possibly you may wish to investigate tissue metabolism. If equipment is available, you might compare the respiratory rates of gill, muscle, and digestive gland of crayfish or of some set of tissues from a different organism. Another line of investigation could lead to the comparison of metabolic rates of a given tissue under different conditions.

B. Feeding and Digestion

The feeding and digestive activity of many animals may be profitably investigated. Some suggestions utilizing two kinds of readily available filter feeders follow.

1. *Daphnia.* Make a smear of petroleum jelly on the bottom of a syracuse watch glass. Add a *Daphnia* in a small drop of water. When the *Daphnia* comes into contact with the petroleum jelly, it will tend to get stuck. The process may be aided by gently pushing the *Daphnia* with a teasing needle. For the study of feeding, the *Daphnia* should be dorsal side down so that its appendages can extend freely between the valves of its exoskeleton. Once *Daphnia* is fastened in the proper position, more water should be added to the preparation.

The feeding mechanism may be observed by adding a suspension of colored material to the water. If a substance such as neutral-red-stained yeast is used, digestion as well as feeding may be observed.

Observe feeding under the microscope. After *Daphnia* has fed for awhile, place it on its side and observe the gut. Neutral red becomes yellow at a pH of 9. Does any region of the filled gut of *Daphnia* change color? How might this color change be related to digestion?

2. *Hydra.* The various species of hydras are filter feeders that use their tentacles as a net to catch prey. The tentacles contain stinging cells called nematocysts.

Place a hydra in a small amount of water in a syracuse watch glass. Observe the hydra under a dissecting microscope. After its tentacles are extended, add several *Daphnia* or brine shrimp larvae (the brine shrimp larvae must be thoroughly washed to remove salt) to the water and observe the feeding activity of hydra. If you watch the hydra long enough, you will see it egest the undigested residue.

The observation of feeding of hydra may be related to the morphology of hydra. For example, the structure of the nematocysts may be determined. Additionally, the general structure of hydra may be examined in a cross section. If you wish to pursue these studies, your laboratory instructor will assist you by providing reference material describing the morphology of hydra and by providing directions for the preparation and study of nematocysts.

C. Temperature

Many animals, such as man and other mammals, have a high degree of independence of environmental temperature. Such independence is expressed in such phenomena as the maintenance of a relatively constant body temperature. However, many kinds of ground squirrels will enter their burrows to avoid midday heat. The body temperature of a bat during its diurnal sleep may decrease to near the temperature of its cave environment.

Have you not adjusted your behavior to cold winds or the hot summer sun?

The activities of most animals are temperature dependent. For many of these, activity is closely correlated with temperature. This relationship is most readily measured by analyzing metabolic rates. Fundamentally, metabolism is a series of chemical reactions. Thus we should look at the effect temperature has on the rates of chemical reactions. The most widely used measure of the effect of temperature on chemical reactions is Q_{10}. Q_{10} is the factor by which the rate of a reaction is increased for a rise in temperature of 10° C. Q_{10} falls between 2 and 3 for chemical reactions. A Q_{10} of 2 is a way of saying that for a 10-degree (C) rise in temperature, the rate of reaction was doubled. However, Q_{10} varies over the temperature range, partly because the rate of enzymatic reactions is not a linear function of temperature. Q_{10} also may vary because of inactivation of some parts of the reaction system due to temperature stress or because an organism has a balanced set of reactions, the net result of which is to reduce the effects of temperature change.

Q_{10} may be calculated for any temperature difference by means of the following formula:

$$Q_{10} = \left(\frac{K^1}{K^2}\right)^{10/(t_1-t_2)}$$

where K^1 is the rate at the high temperature t_1, and K^2 is the rate at the lower temperature t_2. The formula may also be expressed:

$$\log Q_{10} = \frac{10(\log K_1 - \log K_2)}{t_1 - t_2}$$

One possible type of experiment is to investigate the effect of temperature on respiration or on any other activity that can be measured. For example, it is possible to observe the hearts of many animals, such as *Daphnia*, and to count heartbeats in relation to temperature. If you have other ideas, consult with your laboratory instructor about their practicability before progressing further.

Prepare a report describing your investigation. The report should follow, in general, the format presented in Exercise 3, but you may modify that format to whatever extent required, depending upon the nature of your investigation.

SUGGESTED READINGS

Burnett, A. L. and Eisner, T. (1964). "Animal Adaptation," 136 pp. Holt, New York. Paperback.

Schmidt-Nielsen, K. (1964). "Animal Physiology," 118 pp. Prentice-Hall, Englewood Cliffs, New Jersey. Paperback.

EXERCISE

12 EXPERIMENTAL DESIGN

The purpose of this exercise is to permit the student to design and conduct an experiment from minimal information. In the previous exercises the student was introduced to the use of controls in experiments and to the need to choose appropriate materials and methods. In these previous exercises, either a large amount of information was presented to form a foundation for further work or a sample experiment was run to demonstrate the procedure, etc. In this exercise, the student should draw upon ideas, techniques, concepts, etc., presented in the earlier work.

I. EXPERIMENTAL INVESTIGATIONS OF THE RATE OF PHOTOSYNTHESIS

This exercise will use photosynthesis as the functional system to be investigated. The object is to examine some of the environmental factors that affect the rate of photosynthesis.

Photosynthesis may be symbolized by the following input-output equation:

$$6\ CO_2 + 12\ HOH \xrightarrow{\text{light} \rightarrow \text{chlorophyll}} C_6H_{12}O_6 + 6\ HOH + 6\ O_2$$

Your first problem is to determine how to measure the rate of photosynthesis. Examine the input-output equation carefully. At how many steps in the equation is it possible to measure the rate of photosynthesis?

Because of certain practical limitations, we will use aquatic plants in these experiments. The aquatic environment may be more easily controlled and certain measurements of photosynthesis can be made readily in the aquatic system.

The laboratory instructor will discuss with you the advantages and disadvantages of certain methods of measurement. In this way previous experience in the conduct of these experiments will be made available to you so that you will have additional information on which to base your choice of measurement.

The input-output equation also contains some suggestions for experimentation. Obviously, changes in the concentration of CO_2 or the intensity of light ought to affect the rate of photosynthesis. Levels of CO_2 can be varied in aquatic systems in the form of the bicarbonate (HCO_3^-) ion. Light intensity can be varied simply by moving the light source closer to or further away from the plants. The light intensity decreases as the square of the distance. It is not necessary to know the absolute light intensity; arbitrary units may be used. The intensity when the light is at its closest point may be given a value of 100. At greater distances, this arbitrary value will be lowered by the square of the quotient of the distance from the plant to the light at the greater distance divided by the distance from the plant to the light at the closest distance. In other words, 100 (the arbitrary value at the closest distance) is divided by

$$\left(\frac{\text{greater distance}}{\text{closest distance}}\right)^2$$

For example, if the distance from the plant is doubled when the light is moved further away, the light intensity will be reduced to one-fourth. In arbitrary units, the light intensity will be reduced from 100 (at the closest distance) to 25 (at the doubled distance).

Since visible light is part of a spectrum of radiant energy, one may question whether the various wavelengths of light are equally effective in promoting photosynthesis. Wavelength can be varied by the use of filters. A given filter will pass a certain range of wavelengths, but will absorb the rest of the visible spectrum. Thus a red filter passes red light (light of wavelength centering on about 650 millimicrons [mμ]) and absorbs the other wavelengths. If you choose to study the effects of wavelengths of light on the rate of photosynthesis, you may wish also to analyze the role of chlorophyll.

Choose a problem which you wish to investigate. (Because of space, time, and equipment limitations, you may be asked to work in pairs.) Have your choice of problem approved by your laboratory instructor. Once your problem is approved, you are to conduct the experiment and prepare a report according to the directions presented in Exercise 3.

2. ANALYSIS OF THE ROLE OF CHLOROPHYLL

Light, to be effective in photosynthesis, must be absorbed. Light can be absorbed when one or more kinds of molecules are present with the capacity to trap light energy. Two groups of pigments, the chlorophylls and the carotenoids, are normally present in photosynthetic organisms. Thus it seems that one or both of these groups of pigments are primarily responsible for the absorption of light energy. The effectiveness of these pigments as absorbers of light energy may be determined by extracting the pigments and measuring the absorption spectrum in a spectrophotometer. That is, you will be able to determine which part of the visible spectrum is effectively absorbed by the pigment.

A. Determination of the Absorption Spectrum

An algal concentrate or green leaves from other plants will serve as a source of plant pigments. Place the plant material in about 5 ml of 95% ethyl alcohol in a small beaker, and heat to boiling on a hot plate. Remove the beaker from the hot plate and allow the *pigment extract* to cool.

The pigments may be separated by paper chromatography. Paper chromatography operates on the principle that solutes move at different rates when a solvent migrates through a substrate. Filter paper forms a simple substrate through which a solvent may move. Suspend a strip of filter paper from a hook made by opening an ordinary paper clip. One end of the opened clip may be hung from a test tube and the other end acts as a hook on which the strip of filter paper is suspended.

Prepare the chromatographic apparatus as follows:

1. To a clean test tube add carbon tetrachloride (CCl_4) so that it fills the tube to about 1 centimeter (cm) from the *bottom*.
2. The suspended strip of filter paper should just extend into the CCl_4 solvent and should not touch the sides of the test tube.
3. Superimpose a number of horizontal streaks of pigment extract on the filter paper just above the solvent line. Permit the extract to dry between applications.
4. Return the filter paper, suspended on the paper clip, to the test tube with the solvent.
5. Stopper the tube and set it in the dark for 30 min.

During the time the tube is in the dark, the solvent, CCl_4, will migrate up the strip of filter paper. The pigments, which are the solutes, will migrate at different rates. At least three bands of color will be evident after the movement of solvent has occurred. Two green bands are *chlorophylls* and one or more yellow-orange bands are *carotenoids*. Depending on time or inclination, each band of pigment may be examined in the spectrophotometer or the *chlorophylls* and *carotenoids* may be examined as a group.

The filter paper should be removed from the solvent after 30 min and hung in the dark to dry. Then cut the strips of pigment from the filter paper. Place the strips into a beaker containing 3 to 4 ml of benzene. Probably all the pigment strips of each kind from the entire class should be put in a single beaker. All the pigments are readily soluble in benzene.

The laboratory instructor will demonstrate the use of the spectrophotometer. *Do not use the instrument until you have the instructor's approval!*

Decant the benzene into a spectrophotometer tube. Measure the absorption spectrum in a Spectronic 20 or similar spectrophotometer. Because the solvent also absorbs light, a solvent blank (benzene) must be used. Measure absorbancy of light at 25 mμ intervals between 400 and 750 mμ and record in Table I.

TABLE I

Wavelength	Absorbancy	Wavelength	Absorbancy
400		575	
425		600	
450		625	
475		650	
500		675	
525		700	
550		725	

Plot the absorbancy (vertical axis) against wavelength (horizontal axis) to produce the absorption spectrum curve.

B. Testing an Hypothesis

The hypothesis is that chlorophyll absorbs light energy which is used in photosynthesis. If correct, it follows logically that the effects of different wavelengths of light (the action spectrum) on photosynthesis should coincide with the absorption spectrum of chlorophyll.

Compare the action spectrum and the absorption spectrum and discuss whether you would accept or reject the hypothesis.

C. Some Questions

Does temperature interact with light intensity in affecting the rate of photosynthesis?

Do light intensity and bicarbonate concentration interact to affect the rate of photosynthesis?

Photochemical reactions are temperature independent over a wide range of temperature. If you found any effect of temperature on the rate of photosynthesis, how can you account for this effect?

Why is chlorophyll green?

On the basis of their color, would you expect carotenoids to be effective absorbers of light energy in the range of 575 mμ?

SUGGESTED READINGS

Goldsby, R. A. (1967). "Cells and Energy," 115 pp. Macmillan, New York. Paperback.

McElroy, W. D. (1964). "Cell Physiology and Biochemistry," 120 pp. Prentice-Hall, Englewood Cliffs, New Jersey. Paperback.

EXERCISE

13 CELL DIVISION

Growth is characteristic of organisms. In multicellular organisms, growth occurs primarily as a result of the increase in the number of cells. The process of cell division is fundamentally the same in all organisms. Attention is centered on the nucleus where the contained information, in the form of chromosomes, is equally divided between the products of division. Important changes also occur in the cytoplasm, but the morphological patterns associated with nuclear division are easier to follow. The division of the nucleus is mitosis. Mitosis may occur without cytoplasmic division or cytokinesis.

Many important events of mitosis occur before morphological changes are evident. For example, the chromosomes replicate and energy stores are accumulated. The morphological events of mitosis can be considered a mechanism for the orderly distribution of chromosomes to daughter nuclei.

1. MITOSIS

A. The Plant Cell

1. *Meristematic cell, general structure.* The basic plant cell from which others are derived is the *meristematic cell,* a cell that is capable of undergoing division. Locate the meristematic region in the longitudinal section of the onion root tip. The region will look much like Fig. 13-1. Many of the cells will be fixed at some stage of division and several such cells can be seen in the figure. The meaning of these cells will be described later. Many of the cells will be in *interphase;* that is, they will not show evidence of division. Examine one of these cells (such as the cells in the top three rows of cells of

Fig. 13-1) in which the nucleus is present. (Why should the nucleus be missing in some cells?)

Note the location of the nucleus; the size, number, and location of vacuoles; shape of cells; thickness of cell wall; and size of nucleus in relation to the cytoplasm. Contrast this cell with the cell of *Elodea* that you studied previously. In what ways does cell morphology appear to be adapted to cell division?

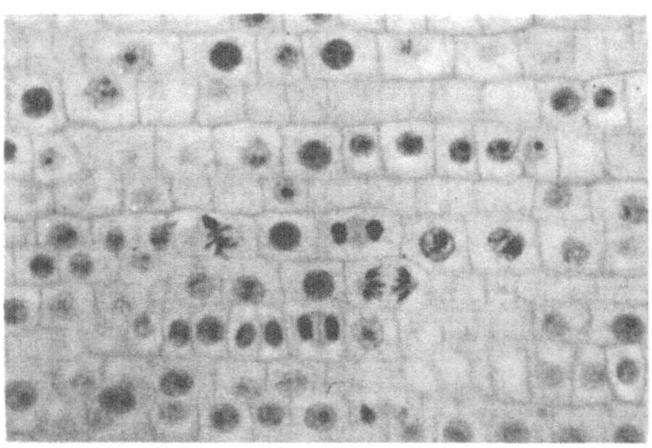

FIG. 13-1. The meristematic region of the onion root tip. Longitudinal section.

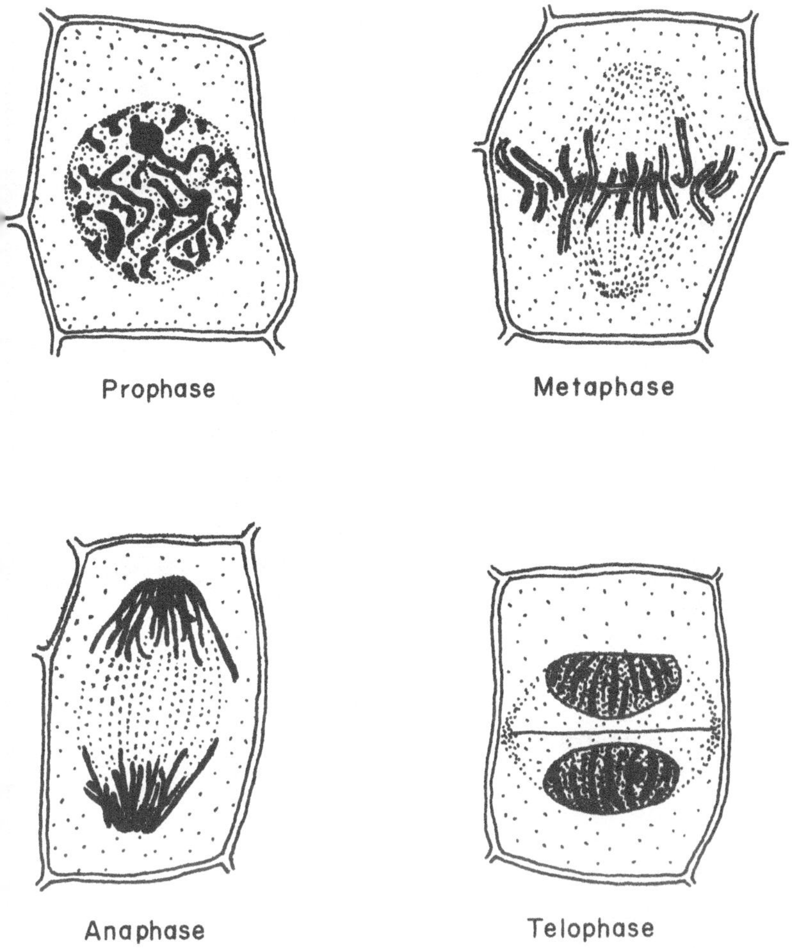

FIG. 13-2.
Mitosis in *Allium* (onion).

2. *Mitosis in the meristematic cell.* After you finish examining the general characteristics of the interphase cell, begin study of mitosis. Although the term interphase implies inactivity, the nucleus is active during this period. Many metabolic activities of the cell are controlled by means of the physiological activities of the nucleus. During interphase, certain physiological events associated with mitosis occur. The proteins of the spindle are synthesized and the chromosomes replicate. However, the chromosomes are not visible as discrete units at this stage because they are elongated and dispersed throughout the nucleus.

The following sequence of phases represents subdivisions of a continuous process. The essential act of mitosis, the replication of the chromosomes, occurs in interphase. The following process may be thought of primarily as a mechanism to distribute the chromosomes equally to the daughter nuclei. With the aid of Fig. 13-2, find each of the following phases of mitosis.

A. PROPHASE. In prophase the *chromosomes* are first evident as long thin threads. They become increasingly visible through a process of shortening and thickening. The slender units are transformed into a coiled structure much as a thin wire can be changed into a coiled spring. The *nucleoli* diminish in size and near the end of prophase disappear. The *nuclear membrane* also disappears in late prophase. The disappearance of the nuclear membrane coincides with the development of the *spindle.* The spindle consists of long-chain protein molecules oriented in a longitudinal direction between two *poles.* Some of the fibers of the spindle attach to a region on the chromosomes known as the *centromere.*

B. METAPHASE. In this stage, the chromosomes become arranged midway between the poles in a region called the equator. The chromosomes are oriented on the equator by means of the *centromere* and its attached spindle fibers. The position of the centromere is visible in the metaphase chromosome by a constriction which is formed because the centromere is oriented on the equator and the arms of the chromosome hang freely and are randomly oriented. The position of the centromere is characteristic for each chromosome and thus divides the chromosome into two arms of varying length.

C. ANAPHASE. The centromeres divide and the daughter chromosomes move slowly to opposite poles. An identical set of chromosomes moves to each pole. The centromeres lead and the arms of the chromosomes trail. The spindle fibers between the poles and the centromeres shorten.

Poleward migration appears to result from the shortening fibers pulling the chromosomes toward the poles. Anaphase ends when the chromosomes become grouped at the two poles.

D. TELOPHASE. Telophase is essentially the reversal of prophase. The spindle fibers disappear, a nuclear membrane forms around each set of chromosomes, the nucleoli reappear, and the chromosomes uncoil to become slender threads.

3. *Cytokinesis.* Evidence of cytokinesis begins to appear at about the time the chromosomes reach their respective poles. A new cell wall forms at the equator. Initially a *cell plate* forms within the spindle, but soon extends to the outer walls to divide the cytoplasm into about equal parts. The spindle then disintegrates and cell division is complete.

4. *Preparation of stained squash of onion root tip cells.* As you learned in a previous exercise, staining is a useful tool for making cellular organelles visible under the ordinary light microscope. For nearly any but the simplest observations, tissue preparation is also necessary. The basic preparative step is fixing the tissue in a solution so that the structures under study are preserved in their natural condition. There is always the problem of how much distortion of the organelles occurs because of fixing; the student of cell structure must work to minimize distortion.

One of the simplest preparations is the squash, which is especially effective for looking at chromosomes. Although some skill must be developed in order to obtain consistently good preparations, care in following directions should result in a satisfactory preparation.

Select an onion root tip from the fixative solution on the demonstration desk. Cut off and discard all but the white glossy tip (2–3 mμ). Place the white tip in 1 or 2 drops of propio-orcein stain on a clean microscope slide and stain for 5 min or less. Then macerate the root tip with a dissecting needle or scalpel and add a cover slip. Macerate the tissue again by pressing carefully on the cover slip with an eraser from a pencil or with a forceps. *Press gently* and *evenly* or the cover slip will be broken.

Examine your preparation and identify as many stages of mitosis as you can.

B. The Animal Cell

1. *Mitosis in the blastula.* Mitosis is readily observed in animal cells in the early stages of embryonic development. Examine the slide of the blastula of the whitefish. (See Fig. 13-3). The behavior of the chromosomes is essentially the same as in the onion cells, except the whitefish chromosomes are more numerous and smaller. The most obvious difference between plant and animal mitosis is in the process of spindle formation. (See Fig. 13-4.)

During prophase one or two radiating structures lie adjacent to the nuclear membrane. Such a structure is the *centrosome* with astral *rays* radiating from it. The centrosome is readily discernible just prior to mitosis and appears as a double structure. The centrosomes separate and migrate along the nuclear membrane until they lie opposite each other. When the nuclear membrane breaks down, they organize the spindle in such a way that the centrosomes are at the poles with the spindle between them. Astral rays also extend into the cytoplasm, but perform no known function.

2. *Cytokinesis.* The process of cytoplasmic division is another point of difference between plant and animal cells. A process of *furrowing,* beginning at the outer edges of the cell in the vicinity of the equator, cleaves the cell in two. The cleavage furrow appears at about the beginning of telophase.

You may find it convenient to label the illustrations of mitosis. Label them sufficiently so that you understand and can describe the sequential steps of nuclear division. Be prepared to compare the processes in plant and animal cells. Can you think of any reason why there should be different methods of cytokinesis in plant and animal cells?

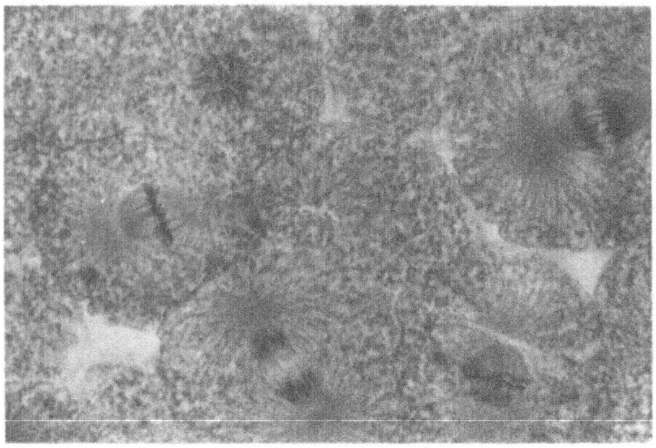

FIG. 13-3. Cells of blastula of the *Coregonus* (whitefish), showing mitosis.

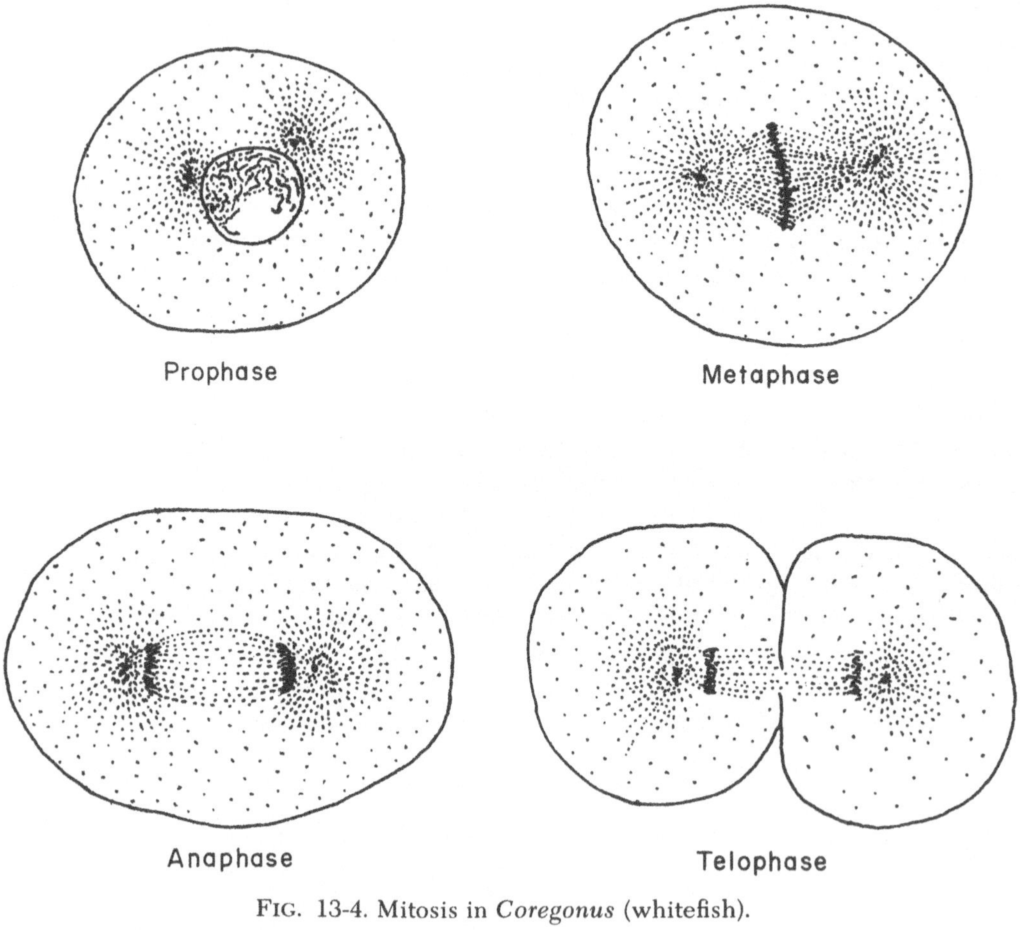

FIG. 13-4. Mitosis in *Coregonus* (whitefish).

C. Time Sequence of Mitosis

The relative length of time required for the completion of each mitotic stage is directly related to the number of cells observed in each stage. That is, the longer a given stage lasts, the more likely it is to be seen. Those stages that last longer will thus be more numerous when the stages are enumerated. If the actual time for the completion of an entire mitotic cycle is known, then the amount of time for each stage may be calculated. If a particular stage makes up 10% of the cells, then that stage would take 10% of the time. If the complete cycle took 16 hours (a reasonable approximation for the onion), then, in our example, the stage would last 1.6 hours. Even though the absolute time may not be known, the relative time may be readily determined. The percent of the total number of cells is also the percent of the total amount of time.

The time sequence of mitosis may be determined by enumerating each stage of mitosis in a large population of cells on the permanent slides of onion root tips. Enumerate the mitotic stages of at least 50 cells and record the number counted and percent of total in Table I.

TABLE I

Mitotic stage	Number counted	Percent of total
Prophase		
Metaphase		
Anaphase		
Telophase		
Interphase		

2. MEIOSIS

At some point in the life cycle of sexually reproducing organisms meiosis occurs. This type of nuclear division results in the chromosome number being reduced by one-half. The reduction of chromosome number from the diploid (2n) to the haploid (n) state is not the essential feature of meiosis, however. Meiosis is a mechanism for generating genetic variability through random assortment of the chromosomes at the first meiotic division. The details of meiosis may be found in any modern biology text and only the essential steps are discussed here.

As in mitosis, chromosomal duplication occurs in interphase. The doubling of the chromosomal number effectively creates a 4n nucleus. Thus, two divisions are required to reduce the 4n state to the n state. Initially, and quite distinct from mitosis, the members of a pair of chromosomes show strong mutual attraction. In the first metaphase, the duplicated pairs of chromosomes lie close to each other to form a tetrad of four chromosome strands. At the first anaphase, each pair of chromosomes separates, one member of the pair going to one cell and the other member of the pair to another cell. This separation of paired (homologous) chromosomes is *segregation.* The way in which one set of homologous chromosomes segregates is independent of the way any other set of homologous chromosomes segregates. Thus random assortment of chromosomes occurs at anaphase of the first meiotic or reduction division. At metaphase of the second meiotic or reduction division, the chromosomes line up in the equatorial plane as they do in mitosis. At anaphase each pair of chromosomes separates; one member of each pair goes to each daughter nucleus.

A. A Simplified Model of Meiosis

A simple model will illustrate the basic principles of meiosis. Your laboratory instructor will provide you with a set of colored toothpicks. The long toothpicks represent one set of homologous chromosomes and the short toothpicks represent a second set of homologous chromosomes. Each member of a pair of chromosomes is of slightly different shade from the other member. This color difference represents the carrying of different specific sets of information on the two members of the homologous pair.

Use the toothpicks to model the chromosome pattern of meiosis. The following events should be modeled by arranging the toothpicks in the correct patterns:

1. Duplication (4n state)
2. First metaphase (tetrad formation)
3. First anaphase (segregation of pairs, 2n state)
4. Second metaphase
5. Second anaphase (n state)

Repeat the process to demonstrate random assortment. How many combinations of chromosomes (length and color) are possible in a nucleus with two homologous pairs? You may wish to write the combinations in the following space and have your laboratory instructor check them for accuracy. If you can manipulate the chromosomes through meiosis to produce all the possible combinations in the haploid (n) nuclei, you should have a good grasp of the fundamental process of meiosis.

To be certain that you understand the difference between meiosis and mitosis, use the same set of toothpicks to model mitosis. Manipulate the toothpicks to demonstrate the following: (1) duplication; (2) metaphase; (3) anaphase.

SUGGESTED READINGS

Alston, R. E. (1967). "Cellular Continuity and Development," 214 pp. Scott, Foresman, Chicago, Illinois. Paperback.

Bonner, D. M. and Mills, S. E. (1964). "Heredity," 115 pp. Prentice-Hall, Englewood Cliffs, New Jersey. Paperback.

Swanson, C. P. (1969). "The Cell," 150 pp. Prentice-Hall, Englewood Cliffs, New Jersey. Paperback.

Wilson, G. B. (1966). "Cell Division and the Mitotic Cycle," 111 pp. Reinhold, New York. Paperback.

EXERCISE

14 GROWTH

Development is a general feature of reproduction that involves the creation of adult form by means of the growth, multiplication, and differentiation of cells. The growing organism is not merely a miscellaneous assemblage of cells, but is composed of cells organized into specific patterns. The establishment of these patterns and the processes whereby the adult form takes shape is known as morphogenesis. One of the central problems of biology is how the processes of development are regulated and organized to produce the integrated organism.

1. PLANT GROWTH

A. Growth as a Response to Environmental Stimuli

Plants respond to stimuli from the external environment primarily by changes in patterns of growth. Such changes are relatively slow and simple at the organismic level, although the cellular mechanisms are complex.

The responses of plants are controlled by hormones. One group of hormones, the auxins, are produced by the growing apices of stems and roots. The auxins migrate from the apex to the zone of elongation where they are specifically required for the process of elongation. If more auxin accumulates on one side of a plant structure than on the opposite side, e.g., in a stem, curvature of the structure results. Thus changes in the direction of plant growth can be interpreted on the basis of an asymmetrical distribution of auxin in the organ involved.

Two stimuli that essentially are constantly acting on plants are gravity and light. Each of these stimuli contributes to the determination of the form of growth and the position in space of plants.

1. *Gravity.* Working in groups of four, prepare the following demonstration of geotropism. Select three germinating seeds of corn which have the primary roots just protruding from the seed. The primary root should not be over 2 mm in length. Line a plastic tumbler with a damp piece of paper toweling. Between the paper toweling and the tumbler insert the seeds about an inch apart and about 2 inches down from the rim of the tumbler. Place the seeds so that the embryo will be next to the tumbler and with the primary root pointing in the following directions:

1. Toward the bottom of the tumbler
2. Toward the top of the tumbler
3. At right angles to 1 and 2

Gently fill the remaining space in the tumbler with moist sphagnum or vermiculite. Add more water whenever necessary.

On the left side in the space below, make a sketch showing the positions of the corn seeds. At the next laboratory meeting, on the right side of the space make a sketch showing the appearance of the corn seedlings.

2. *Light.* Examine the demonstration plants that have been illuminated from one side only and from above. Make a sketch showing the orientation of the plants with respect to the source of illumination. What organs of the plants showed a response?

Examine the demonstration of the two sets of plants, one which was grown under light of high intensity and one which was grown under light of low intensity. List any differences that you observe between the two sets of plants.

Dandelions growing in tall grass have leaves that are nearly erect. However, dandelions growing in a closely mowed lawn have their leaves so arranged that there is a flat rosette against the ground. Explain this difference in growth pattern.

B. Growth as a Response to Hormonal Actions

Growth and organ formation are regulated in plants, and probably in animals, by the interaction of chemical substances which are perhaps best described as hormones. The following experiment illustrates the nature of the effects of hormones.

1. *Effect of gibberellic acid on plant growth.* The gibberellins are naturally occurring hormones which play many roles in the growth and development of plants. One of these roles may be demonstrated by the following experiment. In this experiment, one of the gibberellins, gibberellic acid will be used.

Work together in pairs. Obtain eight test tubes from the instructor. Label these 1, 2, 3, 4, and 1a, 2a, 3a, 4a, respectively. Fill each tube as follows:

Tube number 1, 1a: Solution which contains 0.1 mg of gibberellic acid per liter
2, 2a: Solution which contains 1.0 mg of gibberellic acid per liter
3, 3a: Solution which contains 10.0 mg of gibberellic acid per liter
4, 4a: Distilled water

Cover each tube with a square of parafilm.

Into each tube insert the root of a germinated Alaska pea seedling down through the parafilm into the solution. Make sure the parafilm fits tightly around the root to prevent evaporation. Wrap foil around all the "a" tubes in order to exclude all light from the roots. These tubes should be set aside in the test tube rack provided by the instructor and observed in 2 to 3 days.

Record your results in Table I by measuring the total length of the shoot and the root for each treatment. You might also measure the length between internodes. (The node is the place on the stem from which a leaf arises. An internode is the distance between two nodes.) You could then test the following hypothesis: Giberellic acid causes an increase in the number of nodes.

TABLE I
Mean Growth of Alaska Pea Seedlings

	Light				Dark			
	Distilled water	Gibberellic acid (mg/liter)			Distilled water	Gibberellic acid (mg/liter)		
		0.1	1.0	10.0		0.1	1.0	10.0
Shoot								
Root								

What conclusions can you make concerning the effect of gibberellic acid on plant growth?

In what way(s) might gibberellic acid affect plant cells to produce the observed patterns of growth?

Recent studies have indicated that gibberellic acid may function to increase the effectiveness of indoleacetic acid, an auxin. More indoleacetic acid was present and effective far from the site of application on the experimental plants when gibberellic acid was present than when gibberellic acid was absent.

Can you devise an experiment to see if the growth effects attributed to gibberellic acid are caused by auxin? Outline a procedure and have it checked by your laboratory instructor. If time permits, carry out the experiment. Record and interpret your data in the space below.

C. Plant Meristems and Cell Elongation: The Morphological Basis

In Exercise 13 we examined the structure of the meristematic cell as related to its function of cell multiplication. Meristematic cells are clumped in specific areas of the root and shoot known as *meristems*. These meristems provide for an increase in length and for the development of lateral organs. The major part of extension growth occurs by means of cell elongation. The increase in length of the cell is mainly a consequence of the uptake of water. Concurrent with elongation, but sometimes subsequent to it, differentiation occurs.

1. *The stem apex*. Examine a longitudinal section of the stem tip of *Coleus* (Fig. 14-1) under the low power of the microscope. The cells at or near the

stem tip are meristematic and form an *apical meristem*. Lateral to the tip are leaf primordia developing into immature leaves. Back of the meristematic region are cells that are elongating.

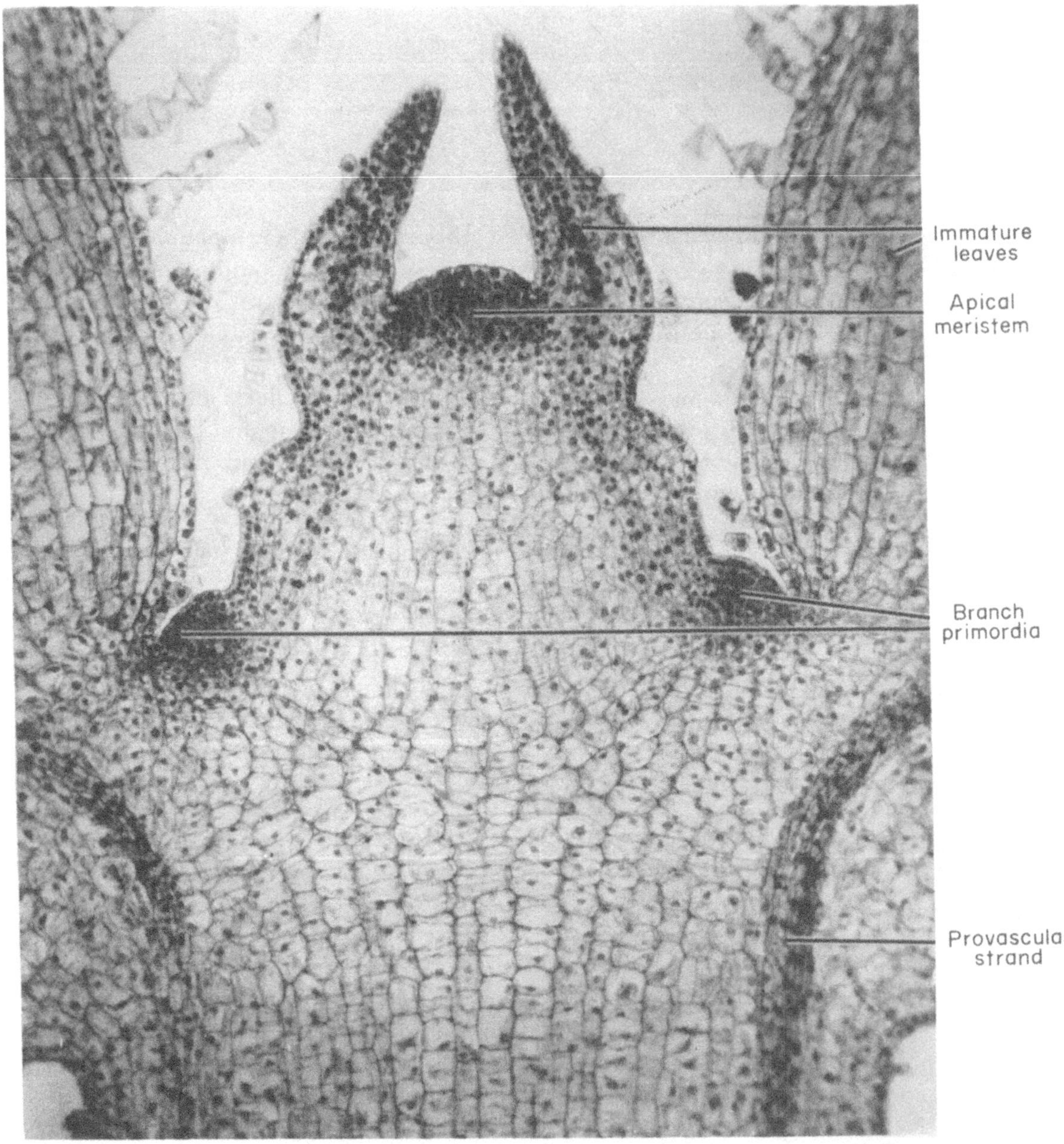

FIG. 14-1. Stem tip of *Coleus*.

Is the region of elongation as clearly set off as it was in the root tip? In the axils of immature leaves are small swellings of meristematic cells that form *branch primordia*. Some of the branch primordia are small and others have developed to the extent that they resemble the central stem tips except for their size. Examine the cells near the base of your section for evidence of differentiation. You may find *provascular strands,* the beginning of vascular tissue, the conductive or transport tissue of plants.

2. *The root apex.* Examine a longitudinal section of the onion root tip. The growing tip can be divided into several regions which are evident in the prepared root tip. At the very tip is the root cap, which acts as a buffer as the tip is pushed through the soil. Behind the root cap is the *meristematic region,* which contains cells that are actively dividing. In the next region, the newly formed cells are increasing in length. In this region of elongation, protuberances of the epidermis, the *root hairs,* are beginning to develop. Behind the region of elongation, the cells gradually mature to form the tissues of the adult organ. When the cells are fully differentiated, they make up the *mature region.*

How does the location of the meristems correlate with the growth responses of plants to the environmental stimuli?

2. ANIMAL GROWTH

The new organism begins life as a single cell, the *zygote,* formed as a result of fertilization. From the zygote a functioning, multicellular organism develops. Development involves three stages: (1) *cleavage* in which the zygote subdivides into a large number of smaller cells; (2) *gastrulation,* in which the cells are organized into basic tissue layers that may be further subdivided; and (3) *organogenesis,* in which the subdivisions undergo further differentiation and organization into discrete, functional entities. Do these processes also occur in plants?

A. Early Embryonic Development of Echinoderms

1. *Prepared slide of starfish development.* The early stages of development in echinoderms, such as starfish, sea urchins, and sea cucumbers, are less complicated than in many other animals and illustrate the basic features of embryogenesis (see Fig. 14-2). Examine a slide that has all of the early stages present. The egg is almost microscopic. The *unfertilized egg* contains a large nucleus with a conspicuous nucleolus and a granular cytoplasm. After the egg is fertilized, a delicate *fertilization membrane,* appearing like a crinkled cellophane wrapper, develops around the egg. The membrane serves to prevent the entrance of additional sperm. Cleavage of the zygote produces a two-celled stage. With each cleavage the number of cells is doubled, producing 4, 8, 16, and 32 cells. The 16-celled stage is called a *morula,* because it has a berrylike form. The 32-celled stage forms a hollow, ball-like structure called the *blastula.* The cavity of the blastula is the *blastocoel.* Note how small the individual cells are by this stage. Continued cleavage produces blastulae with yet smaller cells so that it becomes difficult to see individual cells. When the conditions are appropriate, one side of the blastula pushes inward (*invaginates*) into the blastocoel to form a *gastrula.* The gastrula at the early stage consists of an outer layer of cells, the *ectoderm,* and an inner layer of cells, the *endoderm.* The cavity of the endoderm is the *archenteron* which is open to the outside through the *blastopore.* The archenteron is the first indication of the formation of the gut; the blastopore later forms the anus. The cavity between the endoderm and ectoderm is the remainder of the blastocoel.

2. *Observation of fertilization and early cleavage in sea urchin eggs.* The sexes are separate in the sea urchin, but cannot be distinguished externally. The laboratory instructor will open urchins and strip the eggs and sperm in separate dishes. If time permits, this procedure will be demonstrated to you. The eggs will be collected in sea water and the sperm will be collected dry.

Add *1 drop* of dry sperm to 5 ml of sea water. Allow the diluted sperm mixture to stand for 20 min.

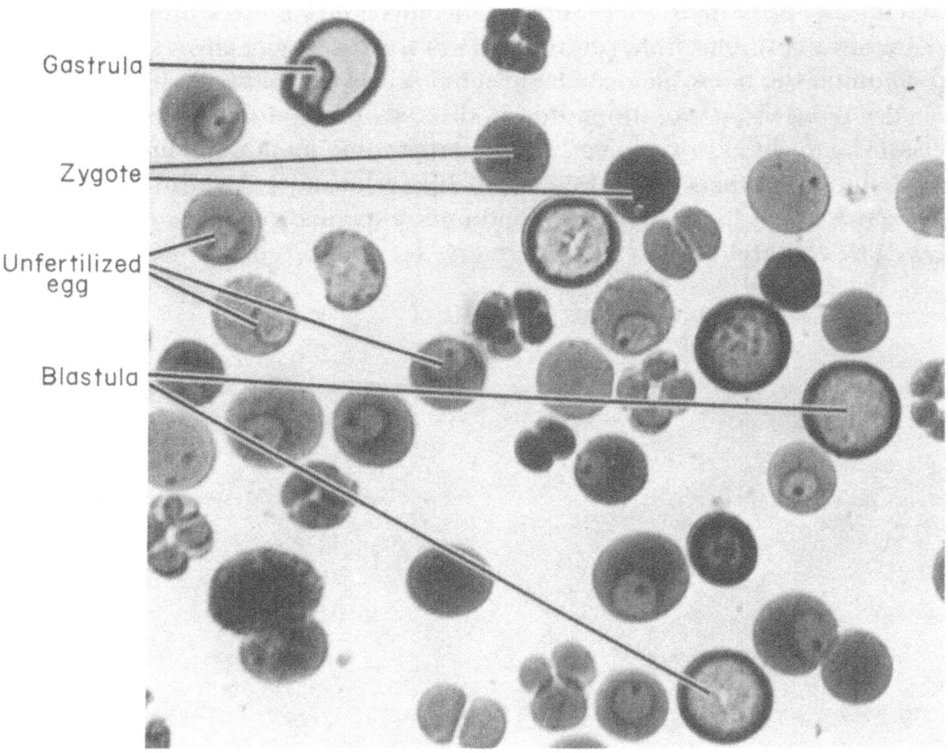

FIG. 14-2. Early embryonic stages of a starfish.

Place a few milliliters of eggs in sea water in a watch glass or small plastic dish. Add *1 drop* of diluted sperm to the dish of eggs and stir.

Immediately transfer several drops of egg-sperm mixture to the concavity on a depression slide and observe under medium power (100× magnification). If fertilization occurs, the fertilization membrane will be found 2–3 min after addition of sperm. The motile sperm will be seen as small motile dots in the solution. The free-living larvae develop in about 24 hours and essentially are specialized gastrulae. You will not be expected to maintain your own mount for 24 hours, but during the lab period you should be able to observe the first or second cleavage. Your laboratory instructor will inform you about how long after fertilization cleavage begins.

Other stages of the living embryo may be obtained from demonstration cultures prepared by the laboratory instructors. Examine this material in a depression slide or on a regular slide on which the cover glass is supported by sand grains. Record your observations.

You may wish to watch fertilization in greater detail. To do so, place a few grains of sand on a microscope slide to support the cover slip. Place a drop of unfertilized eggs in sea water on the slide and cover them with the cover slip. Place the slide under the microscope and adjust to high power. Without alterin the focus or moving the slide, add a drop of diluted sperm at one edge of the cover glass. Carefully look for an egg with a fertilization membrane forming. Note that the membrane arises as a small blisterlike elevation. Make notes of your observations, e.g., describe sperm entrance, especially noting changes in the egg or its activity.

B. Early Embryonic Development of the Frog

The process of cleavage and gastrulation in the frog differ in detail from those of the starfish because of the large amount of stored food or yolk in the egg. We will examine two stages following gastrulation to illustrate the process of subdivision of cell layers into smaller groups that undergo further differentiation.

Examine the series of demonstration slides that illustrate the formation of the early stages of the central nervous system (Fig. 14-3). The *neural groove* first appears as a longitudinal furrow down the middle of the back. The neural groove deepens and its sides grow over and meet above the groove, shutting in a *neural tube* below the surface. While neural tube formation is taking place, a rod of cells is organized between the neural tube and the endoderm (enclosing the archenteron) to form the *notochord,* a supporting structure later replaced by bone. Lateral to the neural tube, masses of cells form the mesoderm, which later grows down between the ectoderm and endoderm and eventually meets below the archenteron. From the endoderm, ectoderm and mesoderm, further subdivision leads to organ formation.

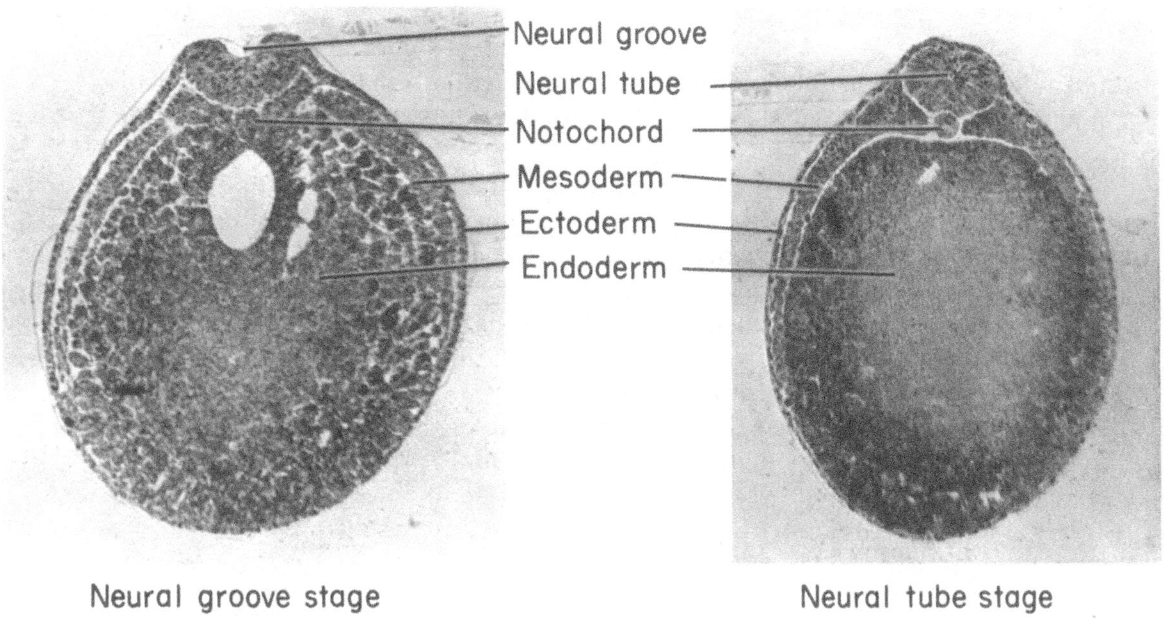

FIG. 14-3. (Left) Neural groove stage of a developing amphibian embryo. (Right) Neural tube stage of a developing amphibian embryo.

3. REGENERATION

Development was presented in the earlier sections as an orderly process of change, change in cell structure or differentiation, change in morphology or morphogenesis. The end result is the integrated organism, which was examined in a previous exercise. Although the cells of the adult organism are characterized by specialization for a particular function, all organisms maintain some undifferentiated or unspecialized cells. The reproductive system is a notable example of the presence of relatively unspecialized cells. Unspecialized cells may function in the adult to regenerate lost or damaged parts. All organisms retain some power of regeneration. Deciduous trees regenerate leaves and flowers annually; a willow stem may regenerate roots and shoots. Some salamanders can regenerate whole legs; man, of course, cannot, and his regeneration is essentially limited to wound healing.

Recent studies have shown that even differentiated cells of some organisms may take part in regeneration. Studies of regeneration are an important means of gaining understanding about the processes of differentiation and morphogenesis and about the nature of organismal organization.

A. Coleus

A fundamental concept of multicellular organization is that a cell functions in a particular way not only because of its structure, but also because of where it is located in the organism.

Cut a *Coleus* stem in half in such a way that you maintain the original orientation of the distal (the free end) and proximal (the end closest to the point of attachment to the plant) ends. Tag each of the cut sections. Label the section furthest from the attached end the "distal section" and label the section closest to the attached end the "proximal" section. Place the two cut surfaces (the two that were formerly adjoining) into a pot of vermiculite or sand or through a parafilm covering of a test tube filled with water. Each stem must have one aerial node and one node planted below the surface. Keep watered and record any evidence of regeneration. Is there any difference in the regeneration of the two sections of *Coleus* stem?

This experiment may be amplified in various ways. Is light necessary for regeneration? Will the plants develop if given only distilled water? Will development proceed more rapidly if mineral nutrients are supplied? Perhaps you can think of other factors to test.

B. Planaria

Planarians are flatworms with a relatively simple multicellular organization and rather remarkable powers of regeneration. Numerous experiments can be performed to test the regenerative capacities of various parts of the organism. Is there a limit to the size of a piece of planarian that can regenerate? Does a head piece regenerate faster or slower than a tail piece? What overall changes in size and shape occur in a regenerating planarian? Select one of the above problems or one of your own to study. Keep careful records of the cuts or sectioning you do. Annotated sketches are probably the best means of recording data. If you section an animal into several pieces, place each piece into a separate dish or in separate compartments in the same dish. Carefully label dishes so that you do not lose track of individual sections.

The worm is most easily observed under a binocular dissecting scope. Allow a worm to extend itself on a moist cork, on a cube of ice, or on a moist microscope slide. Cuts or sectioning should be done carefully with a razor blade. Place the cut worm or sectioned pieces in clean pond or spring water. Cover the dish to retard the loss of water by evaporation. Do *not* feed the worms; they do not "feel" like eating after the operation. The food will foul the water and cause the death and disintegration of the worms or fragments. Keep the dishes in a cool place and away from strong light. Cover dishes lightly to retard evaporation. *Change the water every other day!*

If you make any cuts which are designed to separate two or more portions of a worm without detaching them from the body completely, you will have to renew the cuts frequently the first couple of days until the wounds heal because the cut surfaces tend to fuse back together.

You may wish to collaborate with several of your fellow students on a series of experiments which explore a particular problem. Whatever you decide to do, have your project approved by the laboratory instructor. Record your data in the space allotted.

SUGGESTED READINGS

Barth, L. J. (1964). "Development: Selected Topics," 111 pp. Addison-Wesley, Reading, Massachusetts. Paperback.

Ebert, J. D. (1965). "Interacting Systems in Development," 227 pp. Holt, New York. Paperback.

Saunders, J. W., Jr. (1968). "Animal Morphogenesis," 113 pp. Macmillan, New York. Paperback.

Sussman, M. (1964). "Growth and Development," 116 pp. Prentice-Hall, Englewood Cliffs, New Jersey. Paperback.

Torrey, J. G. (1967). "Development in Flowering Plants," 184 pp. Macmillan, New York. Paperback.

Waddington, C. H. (1966). "Principles of Development and Differentiation," 115 pp. Macmillan, New York. Paperback.

EXERCISE

15 USE OF STATISTICS: MENDELIAN GENETICS

Genetics is the study of heredity, of the mechanism whereby like begets like. However, we know from everyday experinece that, with rare exception, siblings are not identical to each other nor are they identical to their parents. Thus we wish to determine how variation occurs within an overall pattern of replication.

It was the Austrian monk, Gregor Mendel (1822–1884), who formulated the first fundamental postulates of genetics. In 1866, he published his hypothesis of paired factors: i.e., an organism contains two factors for each character (flower color, seed shape, etc.) and each of the paired factors came from a different parent. Mendel formulated two principles concerning the inheritance of paired factors:

1. The two factors in each pair separate or *segregate* in the formation of the germ cells, each germ cell receiving only one factor.
2. The segregation of one pair of factors is independent of the segregation of other pairs of factors: i.e., pairs of factors show *random assortment* in the formation of gametes.

We now know of many exceptions to the "law of independent assortment," but an understanding of this "law" along with Mendel's first principle is fundamental to understanding genetics.

Mendel presumably was ignorant about the behavior of chromosomes during cell division. In the early part of this century several scientists saw a relationship between the behavior of chromosomes and the behavior of Mendel's factors. In 1903, the American cytologist Walter Sutton published the first and fullest analysis of this relationship. In brief, Sutton postulated that Mendel's factors—or genes—are carried in the chromosomes. Such a theory explains that genes occur in pairs because chromosomes occur in pairs; that genes segregate randomly in the formation of gametes because chromosomes do.

1. THE DETERMINATION OF GENETIC RATIOS

The ratios of different genotypes and phenotypes occurring in F_2 generations are known as "Mendelian ratios." When Mendel crossed two red-flowered plants he observed 705 red-flowered and 224 white-flowered plants in the F_2 generation. Thus he obtained about 75.9% red and 24.1% white. Theory states that the percentages should have been 75% and 25% respectively. We would expect to achieve the theoretical percentages if we had an immense population. But, we never have an immense population. Thus, the population with which we work is only a sample of the possible events. The larger the sample, the more likely is the true ratio to be found. Therefore, we are usually satisfied with an approximation of the true ratio. By use of proper statistical procedures, the geneticist can determine the probability that the ratio he found represents the true or expected ratio.

A. CHI-SQUARE

Chi-square (χ^2) is a statistical technique for testing the proportion of individuals possessing a specific trait or character. The chi-square test is used when there are two or more categories into which individuals may be placed. Normally there is an expectation that each category will contain a given proportion of the individuals. But because almost always there is some deviation between the expected distribution and the observed distribution, there must be some test to determine if the deviation from expectation is chance or random deviation.

A chi-square test is carried out as follows:

1. A table is constructed (see Table I) in which the phenotypic combinations form the headings.
2. The observed numbers of each phenotype are placed in the proper columns in the table. (**Note:** Columns are vertical; rows are horizontal in the table.)
3. The expected numbers (devised from a hypothesis concerning genetic theory) are determined and placed in the proper columns in the table. For example; if the expected ratio is 3:1, then the expected row would have $\frac{3}{4}$ of the total number actually counted in the column under that phenotype and $\frac{1}{4}$ of the total number actually counted in the column under that phenotype.
4. The deviations from the expected numbers are determined by subtracting the observed number from the expected number for each column. Deviations may be positive or negative.
5. The deviations are squared and entered in the same column.

TABLE I
Calculating Chi-Square (χ^2)

	1	2	3	4	
Phenotypic combinations	Red flowers	White flowers			TOTAL:
Observed numbers	705	224			929
Expected numbers	696.75	232.25			
Deviations	8.25	8.25			
Deviations squared	68.06	68.06			
Deviations squared, divided by expected numbers	0.097	0.29			SUM: χ^2 0.387

Probability =

6. Each squared deviation is divided by the theoretical number of that column.
7. All the quotients (from 6) are added to give χ^2.
8. The value obtained for χ^2 is converted to a P (probability value from Table II. First, look under the degrees of freedom (df) column and find the df for the problem under analysis. df $= n - 1$, where n is the number of phenotypic combinations. For example, if there are two phenotypes, the df is 1. Second, look along the row opposite the chosen df and find a value of χ^2 closest to the value calculated in 7 above. Most likely the calculated value will not be identical to any value in the table. Therefore, choose the value in the table which is higher than the calculated value. This choice is conservative and will result in a lower probability. Third, choose the probability value at the top of the row in which the selected χ^2 value is located.

One example of this procedure follows. For this example, the numbers obtained by Mendel will be used.

1. The table is prepared for two phenotypes, red flowers and white flowers.
2. There were 705 red flowered plants and 224 white flowers. These numbers are entered in the proper columns in the table.
3. Mendel's hypothesis was that the factor for red flowers was dominant to the factor for white flowers and that his F_1 parents were heterozygous. If segregation occurs, $\frac{3}{4}$ of the offspring should have red flowers

TABLE II
Chi-Square (χ^2)

	Probability										
df[a]	0.99	0.95	0.9	0.8	0.7	0.5	0.3	0.2	0.1	0.05	0.01
1	0.0001	0.003	0.02	0.06	0.15	0.46	1.07	1.64	2.71	3.84	6.63
2	0.02	0.103	0.21	0.45	0.71	1.39	2.41	3.22	4.61	5.99	9.21
3	0.11	0.35	0.58	1.01	1.42	2.37	3.67	4.64	6.25	7.82	11.35

[a] df (degrees of freedom) is equal to $n - 1$; n is the number of phenotypic combinations.

and $\frac{1}{4}$ should have white flowers. There were 929 plants. Therefore $\frac{3}{4} \times 929$ or 696.75 should be red and $\frac{1}{4} \times 929$ or 232.25 should be white.

4. There were more red flowered plants and less white flowered plants than expected. Because the deviations are to be squared, it is not necessary to record the sign (negative or positive).
5. The deviations are squared and entered in the table.
6. Each squared deviation is divided by the theoretical number for that column.
7. The two quotients are summed to give a χ^2 of 0.387.
8. Because the number of phenotypes is two, the degrees of freedom is 1. In Table II in the row opposite a df of 1, 0.46 is the value closest to the calculated χ^2 and is larger. Therefore, the probability value of 0.5 at the top of that row is the value chosen.

Chi-square is a test for a null hypothesis; i.e., there is no difference between the expected numbers and the observed number. By convention, a P value of 0.05 or less indicates a significant chi-square value and the null hypothesis is rejected (the observed numbers do not fit the theoretical ratio). If the P value is greater than 0.05, the null hypothesis is accepted (the observed numbers do fit the theoretical ratios). However, the goodness of fit depends on the probability. One can be relatively certain that the observed numbers do fit the theoretical ratio when the probability is 0.9 or higher. It is possible to have a set of numbers fit two ratios at the same time. For example, determine if 70:30 fits a 3:1 or 2:1 ratio. Thus, we must emphasize the importance of making genetic sense of the ratios and of designing further experiments when the true pattern is not evident. Therefore, in all instances, the P value should be stated along with a statement accepting or rejecting the null hypothesis. The P value tells us that if the experiment is repeated independently many times, we can expect chance deviations as large as or larger than the observed deviations to occur as frequently as the P value. For example, if $P = 0.5$, then deviations as large or larger than the observed deviations could be expected

50% of the time. The higher the probability that deviations occurred by chance, the less likely that some other ratio is the "true" ratio.

B. Inheritance of Seed Color in Corn

1. *Sample size and use of chi-square.* The kernels or seeds are either purple or yellow. If one assumes that the parents were heterozygous, the offspring should evidence a 3:1 ratio. Therefore, the hypothesis that the parents were heterozygous and that one color is dominant and the other recessive may be tested following the procedures described in Part A above.

One question which we wish to explore is how many kernels of corn must be counted to obtain an adequate sample size to test the hypothesis.

Carry out the following procedure: Count the number of purple kernels and number of yellow kernels in one row. Assume an expected ratio of three purple to one yellow. Test the results by chi-square and determine the probability that the deviations could have occurred by chance. Would you accept or reject the genetic hypothesis?

Continue the same procedure, doubling the number of rows counted until all the kernels are counted. Thus, chi-square tests will be made for the distribution of the number of purple and of yellow kernels for 1, 2, 4, 8 and all the rows of kernels. Each count should include all the previous counts. That is, the count for two rows of kernels will include the numbers from the first row plus the numbers from the second row. The count for four rows will include the numbers from the first two rows plus two additional rows, etc. Tables are provided for your use in calculating the chi-square values.

How do the chi-square and probability values change as sample size is increased? Based on this study, what sort of general statement can you make concerning sample size?

Does the chi-square test support the hypothesis that purple is dominant to yellow and that the parents were heterozygous?

2. *Formulation and test of an hypothesis.* Obtain a second ear of corn which also has purple and yellow kernels. Utilizing procedures learned in Part A and B,1 above, determine the number of each phenotype and formulate a hypothesis which can account for the observed distribution of phenotypes. Note that such a hypothesis includes a statement as to the probable genotypes of the parents.

Table for Calculating Chi-Square (χ^2)

Phenotypic combinations	1	2	TOTAL:
Observed numbers			
Expected numbers			
Deviations			
Deviations squared			
Deviations squared, divided by expected numbers			SUM: $= \chi^2$

Probability =

Table for Calculating Chi-Square (χ^2)

Phenotypic combinations	1	2	TOTAL:
Observed numbers			
Expected numbers			
Deviations			
Deviations squared			
Deviations squared, divided by expected numbers			SUM: $= \chi^2$

Probability =

Table for Calculating Chi-Square (χ^2)

Phenotypic combinations	1	2	TOTAL:
Observed numbers			
Expected numbers			
Deviations			
Deviations squared			
Deviations squared, divided by expected numbers			SUM: $= \chi^2$

Probability =

Table for Calculating Chi-Square (χ^2)

Phenotypic combinations	1	2	TOTAL:
Observed numbers			
Expected numbers			
Deviations			
Deviations squared			
Deviations squared, divided by expected numbers			SUM: $= \chi^2$

Probability =

Table for Calculating Chi-Square (χ^2)

Phenotypic combinations	1	2	TOTAL:
Observed numbers			
Expected numbers			
Deviations			
Deviations squared			
Deviations squared, divided by expected numbers			SUM: $= \chi^2$

Probability =

Table for Calculating Chi-Square (χ^2)

Phenotypic combinations	1	2	TOTAL:
Observed numbers			
Expected numbers			
Deviations			
Deviations squared			
Deviations squared, divided by expected numbers			SUM: $= \chi^2$

Probability =

C. Inheritance in *Drosophilia*

Much of our knowledge of modern genetics has come from the study of inheritance in the fruit fly, *Drosophilia melanogaster*. The flies are particularly valuable because large numbers can be readily reared in a short time. The life cycle can be completed in about 15 days at 20° C.

Examine a wild type male and female. Carefully determine the nature of characters such as eye color, wing size and shape, wing venation, and body color. Be certain that you can distinguish the sexes. Use criteria such as genitalia, sex combs, color of posterior, and size. (See Fig. 15-1.)

Each student will be provided with a culture of *Drosophila*. Directions for conducting the analysis (including handling of flies, etc.) of inheritance will be provided at the same time. Tables in which you can record your counts are provided below. The tables are designed so that counts of males and females may be recorded separately. If no evidence of sex linkage occurs, the counts for the two sexes may be summed and placed in the total column. Tables for calculating χ^2 are on p. 144.

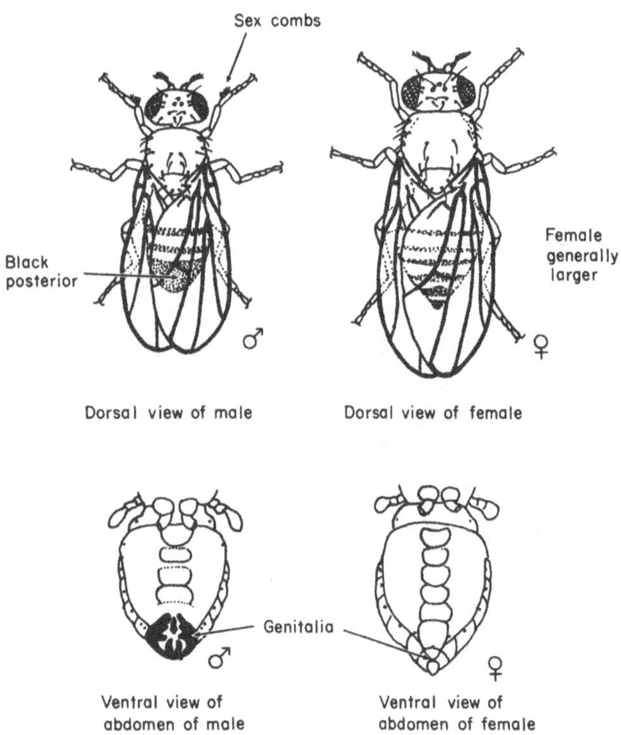

FIG. 15-1. *Drosophila* (fruit fly). (Dorsal view of female is after Dr. T. H. Morgan; ventral view of abdomen of female is after Dr. C. B. Bridges.)

Prepare a written report presenting the results of your investigation. The report should follow procedures previously presented in this exercise and in Exercise 3. Any analysis should include a hypothesis concerning the genotypes of the various phenotypes and of each parental generation; chi-square and probability values of ratios, reasons why linkage, sex-linkage, etc. were or were not present, etc.

F_1 Phenotypes: ______________ ♂ X ______________ ♀
Genotypes: ______________ ♂ X ______________ ♀

F_2 Phenotypes:	Number of flies			
	♂	♀	Total	Ratio
TOTAL:				

F_1 Phenotypes: ______________ ♂ X ______________ ♀
Genotypes: ______________ ♂ X ______________ ♀

F_2 Phenotypes:	Number of flies			
	♂	♀	Total	Ratio
TOTAL:				

Table for Calculating Chi-Square (χ^2)

Phenotypic combinations	1	2	3	4	TOTAL:
Observed numbers					
Expected numbers					
Deviations					
Deviations squared					
Deviations squared, divided by expected numbers					SUM: = χ^2

Probability =

Table for Calculating Chi-Square (χ^2)

Phenotypic combinations	1	2	3	4	TOTAL:
Observed numbers					
Expected numbers					
Deviations					
Deviations squared					
Deviations squared, divided by expected numbers					SUM: = χ^2

Probability =

2. GENETICS PROBLEMS

Fairly reliable ratios are obtained in studies of crosses involving corn and organisms that produce large numbers of young. However, many organisms produce only a few offspring as a result of a single mating. Thus, either the same parents must be mated many times or many similar parents must be mated in order to determine patterns of inheritance. No one mating that produces few young can be expected to show all the possible genetic combinations in the young. Thus we determine the probability or likelihood that a given combination of characters will appear in an individual.

Review the procedure briefly using the following cross. In peas, round seeds (R) are dominant to wrinkled seeds (r) and yellow seeds (Y) are dominant to green seeds (y). Let us cross two plants that are heterozygous round and yellow. Both parents have the same genotype, which can be symbolized as Rr Yy.

First, write the genotype for each parent:

Rr Yy   Rr Yy

Then treat each factor separately, treating it as a simple monohybrid cross. First, write down the possible gametes for roundness or wrinkled for each parent:

R r   R r

Then work out the possible combinations that can occur in fertilization:

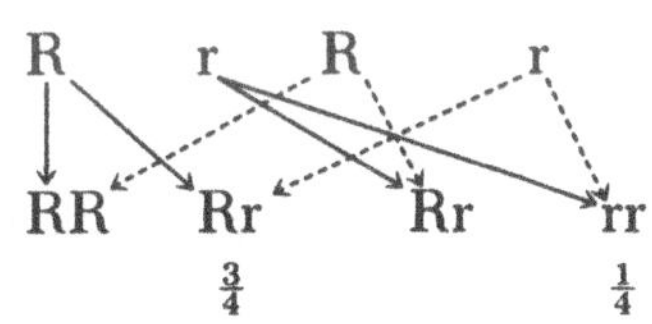

Thus, $\frac{3}{4}$ of the plants can be expected to have round seeds and $\frac{1}{4}$ can be expected to have wrinkled seeds.

Next, follow the same procedure for the color factor. Again we obtain $\frac{3}{4}$ yellow and $\frac{1}{4}$ green. Be certain that *you* can obtain these results before proceeding.

We may now ask the question, how many plants can be expected to have round, yellow seeds? To answer this question, multiply the fraction of plants that might be round times the fraction of plants that might be yellow:

$\frac{3}{4}$	$\times$	$\frac{3}{4}$	$=$	$\frac{9}{16}$
round		yellow		round and yellow

Therefore, any one plant has nine chances in sixteen of being round and yellow. Or if a large population of plants is counted, about $\frac{9}{16}$ of them will be round and yellow.

Using the procedure, work as many of the following problems as necessary until you are certain that you can easily work and understand the basic Mendelian crosses.

A. In peas, seeds may be round (R-dominant) or wrinkled (r-recessive). What proportion of the offspring in the following crosses would be expected to be wrinkled?

1. RR × rr
2. Rr × Rr
3. Rr × rr

B. In peas, seeds may be yellow (Y-dominant) or green (y-recessive). What proportion of the offspring in the following crosses would be expected to be yellow?

1. YY × Yy
2. Yy × Yy
3. yy × yy

C. In peas, the pods may be inflated (I-dominant) or constricted (i-recessive). What proportion of the offspring in the following crosses would be expected to be inflated?

1. II × ii
2. Ii × ii
3. II × II
4. Ii × Ii

D. In peas, the stem length may result in tall (T-dominant) or dwarf (t-recessive) plants. What proportion of the offspring in the following crosses would be expected to be tall? dwarf?

1. TT × tt
2. TT × Tt
3. Tt × Tt
4. tt × Tt

E. If tall peas with inflated pods containing wrinkled seeds are crossed with dwarf peas with inflated pods containing round peas, what possible phenotypes can occur among the offspring?

F. What proportion of the plants (from number E) would be tall with inflated pods and wrinkled seeds from the following crosses?

1. Tt II Rr × tt II rr
2. TT Ii Rr × Tt Ii Rr
3. tt Ii rr × tt ii Rr
4. Tt Ii Rr × Tt Ii Rr

G. In a cross between a tall pea with yellow, round seed (genotype: TT Yy Rr) and a tall pea with yellow, wrinkled seed (genotype: Tt Yy rr), what proportion of the offspring could be expected to be:

1. tall, yellow, round
2. tall, green, round
3. tall, green, wrinkled
4. dwarf, green, round
5. tall, yellow, wrinkled

H. In cattle hornless (H) is dominant to horned (h) and black (B) is dominant to red (b).

1. What proportion of the offspring of the cross B/b H/h × b/b h/h would be expected to be black, hornless? red, horned?

2. With the cross B/b h/h × B/b h/h what is the probability that the first calf will be black, horned? red, horned? red, hornless?

3. What proportion of the offspring of the cross B/b H/h × B/b H/h would be black, horned? black, hornless? red, horned? red, hornless?

I. In addition to the other factors, the gene for white face, or Hereford spotting, (S) is dominant to solid color (s). A cow of genotype B/b H/h S/s is inseminated by a bull of genotype b/b h/h S/s. What is the probability of the calf being black, hornless, and whitefaced? red, horned, and solid color?

In the following problems, genetic reasoning is introduced in addition to or in place of the determination of ratios. The deduction of genetic structure is based on the framework of meiosis and the laws of segregation and random assortment. Also, remember that each offspring gets one-half its chromosomes from one parent and one-half from the other parent. The following problem illustrates both the derivation of ratios and the use of known information in deducing genotypes.

In dogs wire hair (W) is dominant to smooth coat (w):

1. What offspring would result from a cross between a homozygous wire-haired female and a smooth male?

Parents	W/W	w/w
Gametes (each parent produces only one kind)	(W) ↓	(w) ↙
Offspring	W/w wire-haired	

2. If one of the female offspring is backcrossed to her father, what kinds of pups are expected?

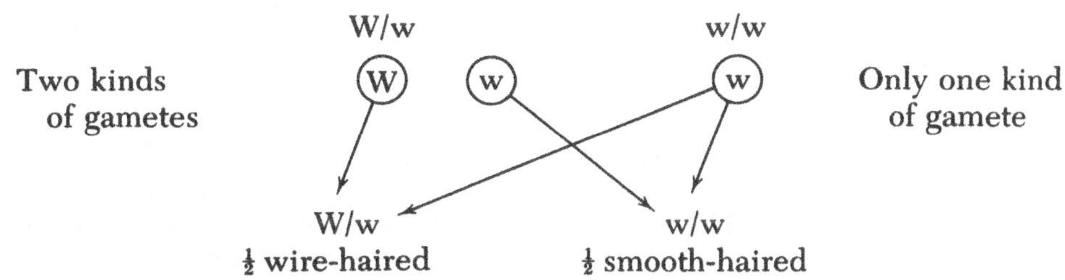

In a large family, approximately 50% would be of each kind. But in a small family almost any combination is possible. Could all the pups be smooth haired in a small family? Could all the pups be wire haired in a small family?

3. If two heterozygotes are crossed, what offspring are expected?

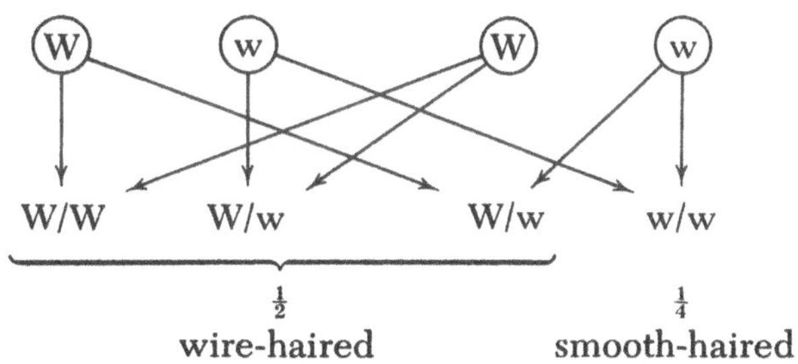

4. A wire-haired male is crossed with a smooth female. Three wire-haired and one smooth-haired pups are produced. What is the genotype of the male?

Writing down the genotypes as far as they can be deduced from the phenotypes:

The male has at least one W W/—
The female must be homozygous w/w

Since one of the pups is W/w, and he must have gotten one of them from each parent, the male must have a w gene. Hence, his genotype is W/w.

1. In man the genotypes D/D and D/d are Rh positive, d/d is Rh negative.

   a. What proportion of the children of two D/d parents would be expected to be Rh+?

   b. An Rh+ man married an Rh− woman and they have an Rh− child. What is the genotype of the man?

   c. An Rh+ man marries an Rh− woman and they have six Rh+ children. Can you be sure what the genotype of the man is? Which is more likely?

d. If the parents in (b) have another child what is the probability that it will be Rh+?

e. Two Rh+ parents have an Rh− child. What are the genotypes of the parents?

2. In the human blood groups, the genotypes are as follows:

group O	i/i	group B	${}_iB/{}_iB$ or ${}_iB/{}_i$
group A	${}_iA/{}_iA$ or ${}_iA/{}_i$	group AB	${}_iA/{}_iB$

a. What kinds of children are not possible from two AB parents?

b. In a case of disputed paternity, the child is group O, the mother A, and putative father is AB. What inference can be drawn from this?

c. What if the putative father were group B?

3. In *Drosophila* the gene vestigial wing (v) is recessive to normal long wing (V), and lobed eye (L) is dominant to normal round eye (l). In a cross between a lobed eye, normal wing and a round eye, vestigial wing, the following offspring were obtained: 92 lobe eye, long wing; 4 lobe eye, vestigial wing; 6 round eye, long wing; 98 round eye, vestigial wing. Determine all the possible genetic information that you can from these figures.

4. In cattle R/R is red, R/R′ is roan, and R′/R′ is white.

   a. What offspring would be expected from a red bull and a white cow?

b. From a red bull and a roan cow?

c. From a roan bull and a roan cow?

d. Starting with a red bull and a white cow, how would you go about getting a white calf?

5. In horses C/C is chestnut, C/W is palomino, and W/W is white.

a. What offspring would be expected from a white stallion and a palomino mare?

b. From a chestnut mare and a white stallion?

c. Starting with a chestnut stallion and a palomino mare, how would you go about breeding a white colt?

6. In cocker spaniels the genotypes and phenotypes are as follows (a dash stands for either a capital or small letter):

A/– B/–	black
A/– b/b	red
a/a B/–	liver
a/a b/b	lemon
A/A B/B	white

a. A red female crossed with a liver-colored male produced a lemon. What were the genotypes of the parents?

b. What proportion of the offspring of the parents would be expected to be black?

7. The difference in Negro and white skin color is inherited as if due to two factors, both lacking in dominance. The genotypes and phenotypes are:

A′/A′ B′/B′	black
A′/A′ B′/B or A′/A B′/B′	dark
A′/A′ B/B, A/A B′/B′, or A′/A B′/B	intermediate
A′/A B/B, A/A B′/B	light
A/A B/B	white

a. What is the darkest colored child that could be produced by a white and a light parent?

b. What is the lightest child that could result from the marriage of a dark and a black person?

c. What range of color could occur among the children of two light parents?

d. What range of color of children could occur among the children of two parents of genotype A′/A B′/B?

8. In man the gene for hemophilia (a disease in which the blood does not clot normally) is recessive and carried on the X chromosome.

   a. What children would be expected from a normal man and a woman heterozygous for hemophilia?

b. From a heterozygous woman and a man with hemophilia?

9. In chickens the male has two X chromosomes, the female an X and Y. Sometimes a hen, because of hormone unbalance, changes into a male but with no change in chromosome makeup. If such a male is crossed with a normal female, what will be the sex ratio in the offspring? (An egg has to have at least one X chromosome in order to hatch.)

SUGGESTED READINGS

Bishop, O. N. (1966). "Statistics for Biology." Houghton-Mifflin, Boston, Massachusetts.

Kalmus, H. (1964). "Genetics," 268 pp. Doubleday, New York. Paperback in The Natural History Library.

Levine, R. P. (1966). "Genetics," 180 pp. Holt, New York. Paperback.

EXERCISE

16 POPULATION GENETICS AND EVOLUTION

Evolution is change. According to modern concepts of evolution, the change occurs as a result of the interplay between the genetic constitution of a population and the environment in which the population is located.

The term population has various meanings. It can refer to the fruit flies within a single culture bottle or it can refer to all the fruit flies in all the culture bottles. The term usually applies to aggregates of interbreeding individuals. Each individual has its genetic characteristics; the population has its characteristics. Ratios are population characteristics. An individual pea plant is either tall or short; the population has some frequency of tallness and shortness. If we assume that any member of the population has an equal chance of mating with any other member, subject to limitation of sex, we can calculate the genotype and phenotype frequencies in the next generation. Actually we have been calculating gene frequency in our study of classical Mendelian ratios. Let us examine such a case and work with gene frequencies.

We can cross a homozygous tall pea plant with a homozygous short or dwarf plant, TT × tt. The total number of genes is equal to unity or 1. If we count the genes, we note that half are T and half are t. Thus, each of these has a frequency of 0.5. Four *possible* matings can occur as shown in Table I. You will note that the table simply expresses the classical Mendelian ratio in the notations of population genetics.

TABLE I

Mating Male × female	Frequencies in mating	Expected frequencies among offspring	
		Genotype	Phenotype
Tall × tall	0.5 × 0.5	0.25 TT	
Tall × dwarf	0.5 × 0.5	0.25 Tt	0.75
Dwarf × tall	0.5 × 0.5	0.25 Tt	
Dwarf × dwarf	0.5 × 0.5	0.25 tt	0.25

The gene frequencies of a population will not change if the population is sufficiently large, if reproduction is random, and if mutation does not occur. This statement of equilibrium of the gene frequencies of a population from generation to generation is known as the Hardy–Weinberg Law. You may work with the example in the above table for as many generations as you like and there will be no change in the gene frequencies. If p represents the frequency of one gene and, q the frequency of the other gene, then $q = 1 - p$ or $p + q = 1$. The gene frequencies resulting from random mating are equal to $(p + q)^2$. The expansion of this expression leads to the equation $p^2 + 2\,pq + q^2$. In the expression, p^2 and q^2 are the homozygotes and $2\,pq$ are the heterozygotes. In our example, $p = 0.5$ (frequency of T) and $q = 0.5$ (frequency of t). Substituting for p and q in the equation, we get:

$$0.5^2 + 2\ (0.5 \times 0.5) + 0.5^2$$

or

$$0.25 + 0.5 + 0.25$$

T Tt tt

Thus, one may calculate gene frequencies using the binomial expression or may use the nonalgebraic method devised for Mendelian ratios.

The use of gene frequencies of 0.5 is useful for the analysis of mechanisms of inheritance. But such frequencies are uncommon in natural populations. For example, the ability to taste phenylthiocarbamide is present as a dominant in 70–80% of the human population in Europe and the Middle East.

1. GENE FREQUENCIES IN HUMAN POPULATIONS

First, we will examine some gene frequencies of characters in human populations. Later, we will examine the consequences of nonrandom reproduction and the subsequent departure from the Hardy-Weinberg equilibrium.

A. Ability To Taste Phenylthiocarbamide (PTC)

Phenylthiocarbamide is an antithyroid drug which prevents the thyroid gland from incorporating iodine into the thyroid hormone. PTC is a derivative of thiourea which many people can taste. This ability to taste PTC is associated with pathology of the thyroid gland. Toxic goiter, an enlargement of the thyroid gland associated with an excessive secretion of thyroid hormone, is more common in tasters than in nontasters, but the excessive form of this disease is more common in nontasters. As with many patterns of human inheritance, the nature of the relationship between "tasting" and disease is unknown.

Taste a piece of paper impregnated with PTC and circle the + if you taste the substance and circle the − in Table II if you do not detect any taste. Record the data for your laboratory section and for the entire class. Can you determine from the phenotype frequencies whether the gene for "taster" is dominant? How could you determine the mode of inheritance?

TABLE II

	Laboratory section		Entire class		Gene frequency
+ −	Number	%	Number	%	
Tasters					
Nontasters					

The gene frequencies may be determined in cases where dominance is known to occur. The homozygous dominant and the heterozygote are usually the same in phenotype and cannot be distinguished. Thus, the easiest genotype to determine is the homozygous recessive in which the phenotype is clearly associated with a known genotype. If q symbolizes the recessive gene, then q^2 represents the frequency of the homozygous recessive, as we learned from the binominal expression. Thus, the square root of q^2 will give us the gene frequency of the recessive. The calculation simply involves the conversion of the percentage of recessive to its decimal equivalent and the determination of the square root. For example, in the cross between homozygous tall plants and homozygous dwarf plants, 25% of the offspring are dwarf. Therefore,

$q^2 = 25\%$, which is converted to 0.25
$q^2 = 0.25$, then $q = (0.25)^{1/2}$
$q = 0.5$

The gene frequency of the dominant, which may be symbolized by p, is $1-q$. Again from our example,

$p = 1 - q$
$p = 1 - 0.5$
$p = 0.5$

Calculate the gene frequencies for tasters and nontasters for the data from the entire class. Do the gene frequencies for your individual laboratory section differ markedly from the class data?

B. Attached or Free Ear Lobe

Free ear lobe is dominant to attached ear lobe. The biological significance of the kind of ear lobe is unknown. Determine which type of ear lobe you have and have your analysis verified by a classmate. Circle the word in Table III which describes your ear lobe type and record the class data.

TABLE III

	Laboratory section		Entire class		Gene frequency
	Number	%	Number	%	
Attached					
Free					

Determine gene frequencies.

C. Ability to Curl Tongue

The ability to curl the tongue in a U-shape is dominant to the inability to do so. Again, the biological significance of tongue curling is unknown. Circle the words which describe your ability and record the class data in Table IV.

TABLE IV

	Laboratory section		Entire class		Gene frequency
	Number	%	Number	%	
Curling of tongue					
Inability to curl					

Determine the gene frequencies.

Is there any similarity in the gene frequencies of these three characters? Do you think that the inheritance of these three characters is a single system or multiple system?

D. Hand Folding

Interlace your fingers and record which thumb is on top. Either your left thumb or your right thumb should be on top. Obviously there are only two ways to fold your hands. Is there any alternative to genetic determination for the manner in which hands are folded? What kind of frequency would you expect if it is a matter of chance which way the hands are folded?

TABLE V

	Laboratory section		Entire class		Gene frequency
	Number	%	Number	%	
Left thumb on top					
Right thumb on top					

Does the frequency distribution support chance determination? Does the frequency distribution fit a 1:1 ratio? How could you test for what ratio the data fit? Do the data permit the determination of gene frequency?

E. Ability To Taste Sodium Benzoate

This chemical is used as a preservative in many foods. To some people sodium benzoate is tasteless. But to others, the substance may be tasted as sour, sweet, bitter, or salty. Taste paper impregnated with sodium benzoate and record your response and the class data in Table VI.

TABLE VI

	Laboratory section		Entire class		Gene frequency
	Number	%	Number	%	
Sour					
Bitter					
Salty					
Sweet					
Tasteless					

Given the distribution of responses, can gene frequency be calculated? If you assume that the ability to taste, regardless of which taste sensation, is inherited as a single gene, can you calculate gene frequency? Do you think such an assumption is warranted? What other kind of information do you need?

F. Frequency of Human Blood Types

The study of blood types also illustrates the principle that the proteins of different individuals tend to be different. This difference is shown by means of an antigen-antibody reaction. Two antigens, called A and B, are found in the erythrocytes. The antibodies, anti-A and anti-B, are found in the serum. Table VII shows the distribution of these antibodies in the four groups.

TABLE VII

Blood groups	Antigen	Antibody
O	None	Anti-A, anti-B
A	A	Anti-B
B	B	Anti-A
AB	AB	None

If an antigen is mixed with its corresponding antibody, an agglutination of the red cells results. The degree of agglutination may depend upon a factor such as the concentration of antibodies in the serum. We shall illustrate the antigen-antibody reaction by determining blood types. Each person will type his/her own blood.

Obtain a clean microscope slide and make two different colored circles with wax pencils. Label one circle anti-A and the other anti-B. Place a small drop of blood in each, following the procedure outlined by your laboratory instructor. Add a drop of anti-A serum, composed of A-antibodies, to the blood in the proper circle and a drop of anti-B serum, composed of B-antibodies to the other. After a few minutes, examine the two circles and note if any reaction has occurred. Fill in Table VIII, place a plus sign where you observe clumping and a minus sign where no reaction occurred.

TABLE VIII

Anti-A reaction	Anti-B reaction	Blood type

If the anti-A reaction only is positive, your blood type is A. If the anti-B reaction only is positive, your blood type is B. If both reactions are positive, you are type AB. If there is no reaction, you are type O. A tabulation of the grouping of each laboratory section and of the entire class will be kept on the chalkboard in the laboratory.

Do the percentages of the individual laboratory sections differ markedly from the percentages for the entire class? Does sample size have any bearing in determining blood type ratios in human populations?

G. Frequency of Red-Green Color Blindness

Deleterious genes are removed from populations by natural selection. Many deleterious genes are maintained only by recurrent mutation. However, we are learning that genes may have mixed effects. One of the best-analyzed examples in human populations is sickle-cell hemoglobin. The individual homozygous for the sickling gene usually dies of anemia before reproductive age. The heterozygous individual shows some sickling and may develop anemia, but also has more resistance to malaria than the homozygous normal. Thus, in malarious areas, the sickle-cell gene may be present in relatively high frequency despite the loss of the gene through the early death of anemics. In this example, the heterozygote has a slightly better chance of survival than either homozygote. Thus the frequency of a gene in a population is probably a result of a balance between its beneficial and its harmful effects.

One gene which occurs in low frequency is the one causing red-green color blindness. It would seem that the bearer of such a gene would be at a slight disadvantage to those who have normal color vision. However, some recent studies have indicated that along with decreased color discrimination, a person red-green color blind may have increased capacity to discriminate shades which appear alike to normal people. Such an individual would have an advantage in seeing other organisms which utilize protective coloration. The advantage of the color blind is not clearly established, but the gene occurs too frequently not to have some advantage. Is it possible that this advantage no longer occurs in modern society? If not, what would you expect to happen to the gene frequency?

Red-green color blindness is determined through use of a color chart. The test is difficult, because it is possible through experience to compensate for the defect, in effect, to learn the response and give it when tested. In order to obtain accurate gene frequencies, each student must be as objective as possible. The test will be given individually so that the "correct" answer cannot influence your reaction to the test. Data for this laboratory section and for the entire class will be recorded and discussed in the next laboratory period.

TABLE IX

Distribution of red-green color blindness		
	Laboratory section	Total class
No. of males positive		
No. of females positive		
No. of males examined		
No. of females examined		
% of males positive		
% of females positive		

Use the pattern of distribution of the presence of red-green color blindness and determine its mode of inheritance. If you have knowledge about the presence of red-green color blindness in your parents, grandparents, brothers or sisters, or children, record this information and use it to verify your analysis (Table IX).

2. NATURAL SELECTION

The action of the environment which changes gene frequencies is natural selection. Natural selection is inexorable. Its necessity was first clearly described by Charles Darwin in *The Origin of Species*. The argument can be summed as follows. Each organism must procure necessary materials for organic maintenance and reproduction and it must have a place in which to live. Space and other resources are limited; therefore the numbers of any species are also limited. But any species has the capacity to increase in numbers geometrically. Thus a population size is quickly reached when only a part of the progeny can survive. The probability of survival will depend, in the long view, on the degree of adaptedness of individuals to the environment in which they live. The degree of adaptedness is determined in part by the genetic constitution. Thus those carriers of genotypes that are slightly better adapted will survive in greater numbers than those carriers of genotypes that are less advantageous. The members of succeeding generations will not be descended equally from all the genotypes of the preceding generations, but relatively more from the better adapted genotypes. Therefore, there will be a change in gene frequencies in the population so that better adapted genotypes will increase and the less well adapted genotypes will decrease. From this discussion we note that evolution is essentially a change in gene frequencies. Evolution occurs because reproduction is nonrandom and the Hardy-Weinberg equilibrium is not maintained.

Natural selection will tend to increase the efficiency by which a population responds to a particular environmental parameter because those individuals that overall are more effective in producing offspring for the next generation will pass on their genotypes. This increased efficiency will continue as long as genetic material is available on which selection can act. Thus organisms may become increasingly better adapted far beyond the degree of adaptation needed for mere survival.

In nature environments usually are variable. Thus variability in a population usually is a positive asset. If sufficient variability is present, there is a strong likelihood that some genetic combination will fit the particular environmental variable at some particular time. In our study of genetics we observed various mechanisms that bring about recombination and hence, variability.

One source of variability is mutation. Mutations of genes occur at various frequencies in populations. In recent years much progress has occurred in the study of mutation and natural selection through the use of microorganisms. Microorganisms are especially useful because they may be cultured under numerous environmental conditions, large numbers of individuals may be obtained, and many generations may be produced in a relatively short time.

A. Natural Selection in *Escherichia coli*

If a culture of the colon bacterium *E. coli* is placed on a medium containing the antibiotic streptomycin, most of the individuals will be killed. However, in a culture of a billion cells, several cells will have mutations endowing the individuals with resistance to streptomycin. Each resistant cell will give rise to a new colony. The number of colonies on the medium may be counted and a mutation rate calculated (number of resistant cell/total number of cells).

1. *Naturally occurring mutants.* Obtain a petri dish which contains brain-heart infusion agar. When your turn arrives, take your petri dish to the front table and obtain a Millipore membrane. Place the membrane, grid up, immediately in the membrane holder in the suction flask. Add 5 ml of buffer (pH 7.0 to 7.5) to the membrane. Add a 1-ml portion of a broth culture of *E. coli* to the buffer, stir, and draw the fluid through the membrane with the suction pump. The bacteria will be caught on the membrane. Using forceps, remove the membrane with its bacteria from the membrane holder and place the membrane on the agar. Place the petri dish in the constant temperature box at 35–37° C. Just before you leave the laboratory, obtain a petri dish with brain-heart infusion agar to which 1 gm/liter of streptomycin was added. Label the dish with your name and laboratory section. Remove your original petri dish from the constant temperature box and transfer the membrane to the streptomycin agar according to the following directions: Rinse your forceps in 70% alcohol. Grasp the edge of the membrane with the forceps, keep the grid up, and place the membrane on the streptomycin agar. Some time between 48–72 hours later, count the number of colonies growing on the membrane. Record your count in Table X. The colonies of *E. coli* will be a gray-white in color. Other bacteria or molds may be present because of contamination. Count only those colonies which, as best you can determine, are *E. coli*.

Dispose of your used petri dishes according to instructions from your laboratory instructor.

2. *Effect of short exposure to ultraviolet radiation.* The number of mutants in a population may be increased by exposing the population to mutagens; e.g., ultraviolet, X-ray, other forms of radiation, and certain chemicals. Bacteria treated with ultraviolet irradiation require a period of growth and synthesis for the induced mutations to become established. This period is referred to as "phenotypic delay," or, more simply, as the time required for the expression of the mutation.

Obtain a petri dish which contains a brain-heart infusion agar. When your turn comes, take your petri dish to the front table and obtain a Millipore membrane. Place the membrane, grid up, immediately in the membrane holder in the suction flask. Add 5 ml of buffer (pH 7.0–7.5) to the membrane. Add a 1-ml portion of a broth culture of *E. coli* to the buffer, stir, and draw the fluid through the membrane with the suction pump. The bacteria will be caught on the membrane. Using forceps which have had the tips rinsed in 70% alcohol, remove the membrane with its bacteria from the membrane holder and place the membrane with the grid up on the agar in your petri dish. Place the half of the petri dish with the agar and membrane in the box under the ultraviolet lamp. Being certain to time the exposure carefully, irradiate the membrane for 10 sec. Then remove the petri dish, cover it with the other half, and place the entire dish in the constant temperature box at 35–37° C.

Just before the end of the laboratory period, obtain a petri dish with brain-heart infusion agar to which 1 gm/liter of streptomycin was added. Label the dish with your name and laboratory section. Remove your original petri dish from the constant temperature box and transfer the membrane to the streptomycin agar according to the following directions. Rinse the tips of your forceps in 70% alcohol. Grasp the edge of the membrane with the forceps, keep the grid up, and place the membrane on the streptomycin agar. Some time between 48–72 hours later, count the number of colonies growing on the membrane. Record your count in Table X. The colonies of *E. coli* will be gray-white in color. Other bacteria or molds may be present because of contamination. Count only those colonies which, as best you can determine, are *E. coli.*

TABLE X

Treatment	Number of colonies of streptomycin-resistant bacteria
A. None	
B. 10 sec of ultraviolet radiation	

Discuss the following hypotheses:

a. Natural selection occurs in the bacterium *E. coli.*
b. Ultraviolet radiation increases the number of favorable mutations.
c. Only favorable mutations are produced by ultraviolet radiation.

SUGGESTED READINGS

Bates, M. (1964). "Man in Nature," 116 pp. Prentice-Hall, Englewood Cliffs, New Jersey. Paperback.

Dunn, L. C. (1967). "Heredity and Evolution in Human Populations," 155 pp. Harvard Univ. Press, Cambridge, Massachusetts.

Ehrlich, P. R. and Holm, R. W. (1963). "The Process of Evolution," 347 pp. McGraw-Hill, New York.

Hamilton, T. H. (1967). "Process and Pattern in Evolution," 118 pp. Macmillan, New York. Paperback.

Smith, J. M. (1958). "The Theory of Evolution," Penguin Books, New York. Paperback.

Wallace, B. and Srb, A. M. (1964). "Adaptation." Prentice-Hall, Englewood Cliffs, New Jersey. Paperback.

EXERCISE

17 POPULATION ECOLOGY

The size of an animal population is determined by diverse factors. The organisms need a place to live, usually called the habitat; are affected by features of the nonliving environment, such as temperature, humidity, salinity, and photoperiod; are affected by their biotic environments, the food they need, their competitors for resources, their predators. Each species has an ability to produce so many young, has a growth rate and a life span. Thus, these physiological attributes of the species interact with environmental factors to determine the growth and maintenance of the population.

1. ENVIRONMENTAL EFFECTS ON POPULATION GROWTH OF *Daphnia*

The influence of some environmental factors on the size and growth of populations of *Daphnia* will be examined in the laboratory. Several species of *Daphnia* are common in ponds or lakes. *Daphnia pulex* is common over much of North America. The larger *Daphnia magna* is less widely distributed but is commonly cultured for laboratory studies. Because species of *Daphnia* are easily handled in the laboratory and because their population biology responds readily to experimentation, we shall use one of these species as our experimental animal. (See Fig. 17-1.)

A. Morphology of *Daphnia*

Examine a mature, living *Daphnia*. A mature female may vary from 1.5 to 2.5 mm in length. The body is divisible into a head and trunk. The head contains a pair of large antennae. The movement of the antennae results in the flealike jumps that are characteristic of the swimming movements. The head also contains a compound eye and a smaller ocellus.

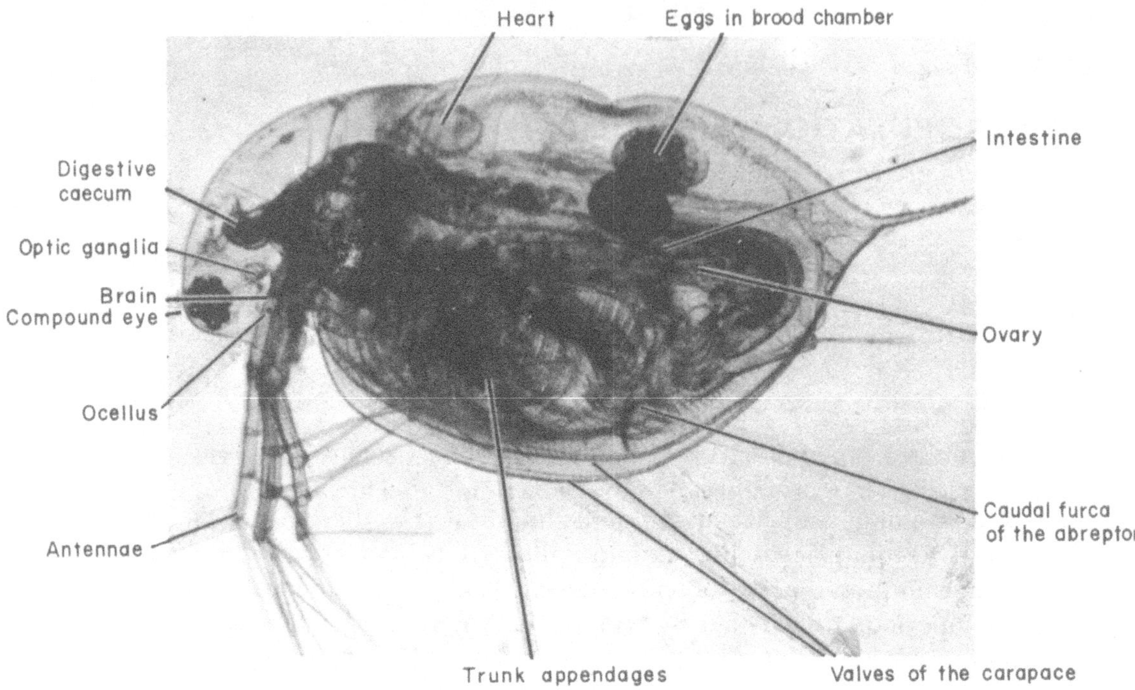

FIG. 17-1. *Daphnia pulex* (water flea). Lateral view.

The trunk is enclosed in a bivalved carapace that is open along the ventral edge and through which the trunk appendages may extend. The trunk terminates in an abreptor which bears a paired row of spines and a clawlike caudal furca. The trunk appendages beat constantly to draw a current of water into the space between the two rows of limbs. Food particles are filtered from the water and swept forward to the mouth. The abreptor may be used to clean out unwanted material from the more anterior parts of the space between the limbs.

Dorsally in the anterior part of the trunk just behind the head is the heart. Posterior to the heart is the brood chamber. Diploid ova produced parthenogenetically pass from the ovary through a narrow oviduct into the brood chamber where they undergo development. The young are released when they have the adult form. When *Daphnia* grows, the exoskeleton is shed so that growth occurs in distinct stages or instars. The first eggs are laid in the fifth, sixth, or seventh instar.

Under some "unfavorable" conditions, some of the parthenogenetically reproducing individuals produce a brood of males, or, rarely, a mixed brood of males and females. Other haploid eggs are formed and fertilization occurs.

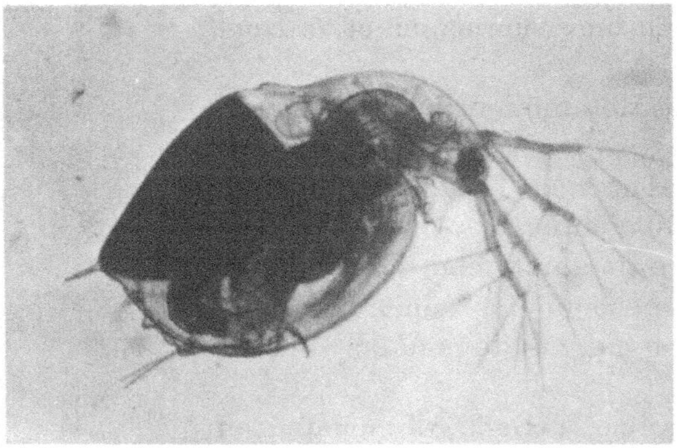

FIG. 17-2. *Daphnia pulex* with an ephippium.

The fertilized egg develops into a larger and darker egg. Also, the carapace around the brood pouch undergoes darkening and thickening to form a protective case known as an ephippium because of its general resemblance to a saddle. At the next molt, the eggs and ephippium are shed as a unit. The eggs are arrested in an early stage of development and will lie dormant until "favorable" conditions develop (Fig. 17-2).

B. Population Experiments

The population experiments will be designed and conducted by the entire laboratory section. Several experiments may be conducted by the class and if the class is large enough, the students should form small groups. Each group may organize and conduct one experiment.

Some suggested experiments follow:

1. Effect of temperature
2. Effect of photoperiod
3. Effect of light intensity
4. Effect of light quality
5. Effect of initial population size
6. Effect of quantity of food

State your experiment in the form of a hypothesis. Clearly describe the materials and methods to be used. You do not need to repeat the general procedure for culturing *Daphnia* described below. Each group should choose a chairman who will be responsible for collating the data gathered by each member of the group. Each student will write a research report according to instructions in Exercise 3.

1. *General procedure.* All students will follow the same basic procedure.

   a. Place 100 ml of aerated culture medium into your container (or containers).
   b. Add the proper number of *Daphnia* as determined by your experimental design.
   c. Change the culture medium three times a week. The changes should occur in two 2-day intervals and one 3-day interval.
   d. Immediately after changing the culture medium, feed the *Daphnia*. You will be provided with an algal suspension of a standard concentration for feeding. The amount of algae fed will depend on your experimental design.
   e. Count all the *Daphnia* in your culture once a week. All members of the laboratory section should count on the same day of the week and the same day of the week should be used throughout the course of the experiment. Remove all dead animals and also record the number of animals with an ephippium. Any loose ephippia should also be counted and recorded.

2. *Specific procedures.* All stated procedures will depend on the particular experiments. As stated above, these procedures should be clearly described in your written report. The following space should be used for constructing a table in which your data are recorded.

2. EFFECTS OF CULTURE MEDIUM ON POPULATION GROWTH AND SURVIVAL OF *DAPHNIA*

The medium in which the *Daphnia* are cultured is an artificial one and balanced so that the various ions are present in sufficient amounts and in proper relationship to each other so that the growth and function of the organisms is as normal as possible.

The medium consists of each of the following compounds added to one liter of glass distilled water:

Amount	Compound
52 mg	$MgSO4 \cdot 7\ H_2O$
75.6 mg	$NaHCO_3$
21.9 mg	$CaCl_2 \cdot 2\ H_2O$
1.0 mg	K_2HPO_4
6.4 mg	$KHCO_3$
18.1 mg	$Na_2SiO_3 \cdot 9\ H_2O$
0.15 mg	Fe (as ferric citrate)

Variation in the amounts of ions or of elements in the ions is another environmental variable. The effects of various ions are not only of theoretical interest, but of great practical importance because of the great amount of substances dumped into our natural waters by industries.

A. Some Suggested Problems

Before any problem is explored, you should have some idea of why these various compounds are used. You should also have some idea why various elements are present. For example, why are iron (Fe) and magnesium (Mg) supplied?

1. Vary the concentration of one compound while others are held constant.
2. Eliminate one compound (but maintain the osmotic concentration of the entire medium in a range favorable for growth).
3. Increase or decrease the concentration of each substance by the same amount so that the net effect is to change the osmotic concentration of the medium, but not the proportions of the constituents.
4. Vary ionic ratios of such ions as K^+/Na^+, Ca^{++}/Mg^{++}, $Ca^{++} + Mg^{++}/Na^+ + K^+$.

Any of these experiments, or one of your own choosing, will require considerable time and care to conduct, but can yield interesting results. If you attempt one of these experiments, try and relate the experiment and its results to the environment in which *Daphnia* lives.

SUGGESTED READINGS

Andrewartha, H. G. (1961). "Introduction to the Study of Animal Populations," 281 pp. Univ. of Chicago Press, Chicago, Illinois. Paperback.

Boughey, A. S. (1968). "Ecology of Populations," 135 pp. Macmillan, New York. Paperback.

Kormondy, E. J. (1969). "Concepts of Ecology," 209 pp. Prentice-Hall, Englewood Cliffs, New Jersey. Paperback.

EXERCISE

18 RESPONSIVENESS

Responsiveness is a primary characteristic of life. The environment constantly stimulates an organism; in effect, the steady state is constantly stressed. The reaction of the organism to maintain the steady state is the response. In general, an organism responds to environmental stimuli that have meaning for the organism; that is, an organism must be sensitive to those stimuli that have essential effects. Thus, many organisms develop receptor systems tuned to the essential stimuli and nonessential environmental changes are ignored.

Any response involving movement of the organism is behavior. The complexity of behavior varies greatly among organisms. Many responses are innate; that is, a particular stimulus evokes a particular response. Many organisms, e.g., animals, can modify their responses with experience; that is, they can earn.

Some of the innate responses involve orientation to particular stimuli. A *tropism* is an orientation to a stimulus that occurs in place, that is, there is movement by the organism but not locomotion. A kinesis is an undirected locomotory response, i.e., the direction of the locomotion has no correlation to the direction or location of the stimulus. In kinesis, the frequency of turnings or the speed of locomotion depends on the intensity of stimulation. A taxis is a directed locomotory response. A tropism or a taxis may be positive (the organism orients toward the stimulus) or negative (the organism orients away from the stimulus).

The responses of organisms are integrated by hormones or by the nervous system or by both. Responses mediated by hormones are slow, taking hours or days to occur. Responses mediated by the nervous system are rapid and may occur in seconds. Responses are carried out by effectors. Some animal effectors are examined in Exercise 10.

1. RESPONSIVENESS IN PLANTS

A. Effection

Effection in plants occurs primarily by changes in growth. These changes were discussed in Exercise 14.

B. Turgor

Plants also respond to stimuli by means of changes in turgor. Such a response occurs in the sensitive plant, *Mimosa pudica*. Special effectors, the pulvini (singular: pulvinus), appear as slightly swollen green structures at the base of each primary and secondary leaflet and at the base of the petiole of each leaf. The pulvinus consists of a mass of parenchyma tissue surrounding the vascular bundle.

Examine a plant and note the size and location of the pulvini. Record below the responses of the plant to a series of stimuli initiated by the laboratory instructor.

2. RESPONSIVENESS IN ANIMALS

The development of a complex system of receptors, effectors, and modulators in animals results in more complex responses at the organismic level than those responses observed in plants. Responses may be immediate and may involve locomotion; e.g., taxis and kinesis. However, sessile animals may show tropistic movements.

A. Light

1. *Planarians.* These primitive worms are especially favorable for the study of responses to light because their photoreceptors are simple and the variety

of response is limited. Because sensory adaptation to light can occur, the planarians are kept in the dark between experiments.

a. Using a lateral source of light, observe the creeping of planarians. Shift the light source 90° and observe the turning reaction. Is the turning response one of "trail and error" or is it a directed response? How does this response compare with that of the fly larva observed in Exercise 1? Record your observations.

b. Place a dish which has one half of the bottom white and the other half of the bottom black under an overhead light. Place ten planarians in the center of the dish and leave them undisturbed for 1 hour. Then count the number that have come to rest on the dark and on the light backgrounds. Record your results in the table on the chalkboard and in Table I in the row "individual results." After all individual results are recorded, copy them into your table and sum them to obtain the total class results.

TABLE I

	Number on white	Number on black	Percent on black
Individual results			
Total class results			

Is there any difference in the number of animals on the white and black backgrounds? Was the difference the same in all individual tests? Calculate the percent of animals on black for each individual test and for the class totals. What comments can you make concerning the advisability of using large numbers of animals or of repeating experiments? Instead of calculating the percent on black, the data may be tested by chi-square.

Several other problems may be investigated. For example:

> Do white planarians respond in the background experiment in the same manner as the dark planarians?
> Does light intensity have any effect on this response of the planarians?
> Does the kind of light (red, green, blue) affect the response of the planarians?

State a problem (one of the above or one of your own) in the form of a hypothesis. Conduct an experiment to test the hypothesis. Record your methods, data, and discussion in the space allotted. Remember that exposure to light before the experiment is conducted may influence the responses and should be carefully controlled.

2. *Daphnia.* Many free-swimming aquatic animals have a cycle of daily vertical migration. By day the animals descend to greater depths and by night they rise. There are many variations in the amount and timing of the movements, but one pattern occurs nearly universally. The animals are closer to the surface at night and deeper by day. Such a pattern suggests a reaction to light. The response could be to the presence or absence of light or to the intensity of light. Many, if not all, species of *Daphnia* respond to light and some undergo vertical migration.

The responses of *Daphnia* to light are more readily studied in a tall cylinder. A graduated cylinder, of at least 100 ml, is a convenient container because the column of water is divided into segments and the number of *Daphnia* in each segment can be easily counted.

Use at least 20 dark-adapted *Daphnia* in 100 ml of aerated pond water or *Daphnia* culture medium. There are several experiments that you might attempt.

A. INTENSITY. What is the pattern of distribution of *Daphnia* under light of high and low intensity? Is there a different distribution of *Daphnia* in the cylinder when it is lighted from the top as compared to lighting from the side? Does temperature affect the response of *Daphnia* to light? Do light-adapted and dark-adapted *Daphnia* respond differently?

B. COLOR. Describe the behavior of *Daphnia* under red and blue light. Do starved and well-fed *Daphnia* respond in the same fashion? Is any other color of light effective?

Any of the problems suggested in these questions may be restated as a hypothesis. State below a hypothesis which you will test. Record your methods, data, and discuss whether the data support the hypothesis.

B. Light and Humidity

1. *Pill bug.* The pill bug or sow bug is a terrestrial crustacean that is unable to withstand prolonged exposure to dry air. We, then, may ask the question, does the pill bug "know" if it is in dry air. Or do other stimuli, such as light, direct the animal to the proper environment? The following experiments will test the responses of these animals to light and humidity.

   a. Fit a piece of *slightly damp* filter paper or paper toweling into the small half of a petri dish. Add five pill bugs and cover with the large half of the petri dish, half of which was painted black to omit light. Place the dish under the light and observe and record over a period of time the activities of the pill bugs. Do the animals respond to light? What happens when the filter paper drys out? Interpret your observations.

   b. Place cotton well soaked in water in one end of a plastic tube and place a desiccant, such as anhydrous calcium chloride, in the other end. After each end is carefully plugged, let the tube lie for 30 min. A gradient of humidity will extend from the wet end to the dry end of the tube.

After 30 min, remove the center cork and quickly add eight pill bugs through the central opening. Place the tube in the dark. After 30 min, determine the location of each animal in the tube and mark its position on the following chart. Is there any marked orientation of the animals in the tube? Considering the nature of the response would you expect to find all pill bugs aggregated in one end or the other?

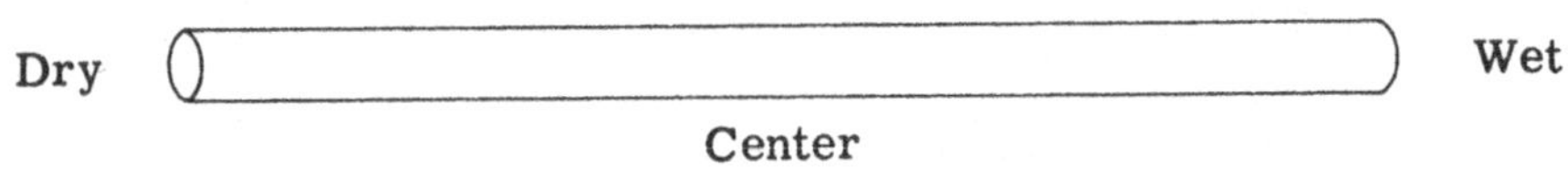

Interpret your results.

Can you construct a model of behavior for the pill bug which will explain how its behavior is directed toward survival of the organism? Base your model on the results of these experiments.

2. *Tribolium or Tenebrio.* These two beetles, the flour beetle and mealworm, respectively, live in environments much more dry than those of the pill bug. Test the responses to light and humidity of either or both of these insects. Design your own experiment. Can you design a system that will permit the investigation of light and humidity simultaneously?

3. ADDITIONAL EXPERIMENTS

In the first laboratory you made some qualitative observations on the behavior of a fish and of a fly larva. Since that time you have learned to analyze biological phenomena quantitatively. At this time you have the opportunity to refer to your earlier descriptions and to consider ways in which you can analyze the conditions under which certain responses occurred.

SUGGESTED READINGS

Davis, D. E. (1966). "Integral Animal Behavior," 118 pp. Macmillan, New York. Paperback.

Dethier, V. G. and Stellar, E. (1964). "Animal Behavior," 118 pp. Prentice-Hall, Englewood Cliffs, New Jersey. Paperback.

Fraenkel, G. S. and Gunn, D. L. (1961). "The Orientation of Animals," 376 pp. Dover, New York. Paperback.

Johnsgard, P. A. (1967). "Animal Behavior," 156 pp. Brown, Dubuque, Iowa. Paperback.

Tavolga, W. N. (1969). "Principles of Animal Behavior," 143 pp. Harper and Row, New York. Paperback.

Van der Kloot, W. G. (1968). "Behavior," 166 pp. Holt, New York. Paperback.

EXERCISE

19 RESULTS OF EVOLUTION: BIOTIC DIVERSITY

It is uncertain how many species of organisms inhabit the earth. Conservative estimates indicate about 350,000 species of plants and 1,200,000 species of animals. Such an array of organisms cannot be studied without some means of systematizing the information. Simply arranging the species into a coherent scheme is a tremendous task.

Man has long attempted to bring order into biotic diversity. One of the first attempts was by Aristotle (384–322 B.C.). He categorized animals on the basis of certain attributes: those with or without red blood; viviparous or oviparous; with perfect eggs or imperfect eggs. Aristotle did not construct a formal system of classification. First attempts at formal classification came in the sixteenth century. Two problems occupied the systematists. First, some system was needed for naming the various species of organisms and, second, some means of grouping was needed. Efforts to solve these problems culminated in the work of Linnaeus (1707–1778). Linnaeus developed the binomial system; that is, each species is given two names, the first name being designated the generic name and the second name being designated the species name. Latin was chosen as the language of classification because it is a language that is stable and therefore universal. For example, the name of man is *Homo sapiens.* Note that the generic name is capitalized and the species name begins with a lower case letter. Scientific names should be printed in italics or underlined.

In addition, Linnaeus devised a system of grouping. He introduced the categories of order, class, and kingdom. A number of similar species were placed in a genus; similar genera were placed in an order; similar orders were placed in a class and similar classes were placed in a kingdom. Thus, classification took on a hierarchial arrangement. Linnaeus believed that species were immutable, having been fixed by special creation. Thus, all one had to do was to study organisms carefully to perceive their natural relationships and place them into their proper "pigeonholes."

The concept of species changed with the development of evolutionary thought following the publication in 1859 of *The Origin of Species* by Charles Darwin. Species are no longer considered immutable and it is not always possible to clearly segregate populations into species. The other categories of the classification system likewise are not discrete units but represent an attempt to draw boundaries around an essentially continuous array of biological organization. In some cases, a particular grouping is fairly discrete; in other cases, the grouping is arbitrary. However, the system of grouping has persisted and is useful for systematizing knowledge.

Each category of the system of grouping is a taxon (pl. taxa). Today there are seven primary taxa, presented here to illustrate their heirarchial arrangement.

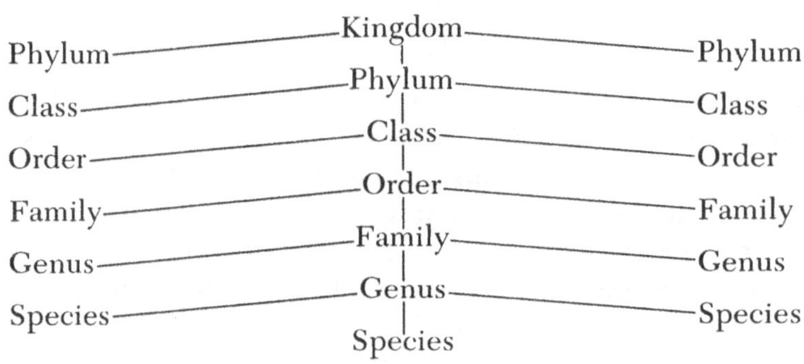

In the following list, the classification of two organisms is presented (**Note:** In plant taxonomy, the taxon "Division" may be used instead of "Phylum"):

Kingdom	Animalia	Plantae
Phylum	Chordata	Tracheophyta
Class	Mammalia	Angiospermae
Order	Primates	Fagales
Family	Hominidae	Fagaceae
Genus	*Homo*	*Quercus*
Species	*sapiens*	*alba*
Common name	man	white oak

In the laboratory work to follow we shall examine some of the patterns of organization of plants and animals. Recent advances in knowledge of many groups of organisms have indicated that an attempt to divide the biotic world into two kingdoms is unrealistic. Rather, there appear to be four major divisions, the Monera, Protista, Metaphyta, and Metazoa. The Monera include the bacteria and blue-green "algae" and are distinguished from all other organisms in that they do not have a true nucleus. The Protista includes an enormous assemblage of organisms, including many that have characteristics of both plants and animals. The Protista are distinguished by a basic acellularity or unicellularity, although some members of this group are multicellular. The Metaphyta include all the advanced multicellular plants and are distinguished from the Protista by the presence of multicellular sex organs and the presence of more or less distinct embryonic stages. The Metazoa are also multicellular. They are distinguished from the Protista in that their cells are organized into tissues and usually into organs or organ systems. They also have embryonic, and usually larval, stages in development.

1. MONERA

A. Bacteria

The bacteria contain the smallest known free-living cells. The cells may be spheres, called cocci, rods, called bacilli, or coiled, called spirilla. Bacteria usually have cell walls of chitin or some other complex polysaccharide. Reserve foods are stored as glycogen in most bacteria. Some bacteria are autotrophic, being chemosynthesizers or photosynthesizers. Most bacteria are heterotrophic. In the ecosystem, bacteria make up the trophic level called the decomposers. They are especially important in the cycles of materials as bacteria convert fixed nitrogen and carbon in forms available to the ecosystem.

Bacteria are economically important in various ways. Many cause diseases of man and his domesticated animals and cultivated plants. Many are important in fermentation in the production of cheese and vinegar. Still other bacteria produce antibiotics, such as streptomycin and aureomycin. Examine the demonstration material of various bacteria.

B. Blue-Green Algae

These organisms probably are not algae in the same sense as are the algae we will examine later. These organisms are usually photosynthetic, containing chlorophyll a. In addition, they contain other pigments, some of which are found in no other group of organisms. The cells lack aqueous vacuoles and store a unique type of starch. Chloroplasts are lacking. Sexual reproduction is unknown in this group.

Most blue-green forms occur in fresh water. They characteristically grow in colonial filaments, often greatly branched, or in clumps of various shapes and frequently are embedded in a secreted gelatinous material. Most blue-greens are producers. They may not form an important direct source of food for aquatic animals, but are most likely utilized as detritus. They are important as oxygenators of aquatic environments. In addition, many of them are able to fix atmospheric nitrogen. Examine *Anabaena* and other demonstrations of blue-greens.

2. PROTISTA

A. The Algae

The algae as a group may be considered as a level of organization characterized by possessing unicellular sex organs or multicellular ones in which every cell forms a gamete. The various groups of algae may be distinguished by certain biochemical features and specific structural features.

Most algae are aquatic, both marine and fresh water. They are especially important in the ecosystem as producers and many are utilized directly by animals. Man uses more than 70 species of red algae as food.

Three fundamental patterns of reproductive or life cycle occur in the algae (and in all organisms):

1. Plant (n) → gametes → zygote (2n) → meosis → plant (n)
2. Plant (2n) → meiosis → gametes (n) → zygote (2n) → plant
3. Plant (2n) → meiosis → spores (n) → plant (n) → gametes (n) → zygote (2n) → plant (2n)

In these patterns the plant may be n or 2n so that meiosis may occur in the formation of gametes or may occur in the zygote. Spores may be formed asexually in many algae and in some a regular alternation of generations occurs. That is, a sexually reproducing plant produces gametes that give rise to a plant that reproduces asexually to form spores and the spores develop into a plant that will reproduce sexually.

1. *Green algae.* Examine the unicellular, motile alga *Chlamydomonas*. The single chloroplast fills most of the cell. The nucleus may be located centrally. Two flagella provide locomotion.

Spirogyra is a filamentous alga in which meiosis occurs after fertilization. Sexual reproduction occurs by means of *conjugation* in which a protoplasmic bridge forms between two cells of two plants and the contents of one cell flow into the adjoining cell. The fused double cell secretes a heavy wall around itself to form a *zygospore*. When conditions are right, the zygospore develops into a new filament.

Ulothrix is an example of an alga that may form motile asexual spores known as *zoospores*. Each zoospore develops directly into a new plant without fertilization. Sexual reproduction also occurs, but does not alternate with spore formation.

Cladophora exemplifies alternation of generations. Both generations are identical in structure except for chromosome number. Examine other demonstrations of green algae.

2. *Brown algae.* The brown algae are almost exclusively marine. *Fucus* has a size and complexity of organization that exceeds those of other algae and many land plants. In *Fucus*, meiosis occurs in the formation of gametes. Ob-

serve a plant of *Fucus* and note the location of the reproductive organs. Observe other examples of brown algae.

3. *Red algae.* The red algae are also mainly marine and have some of the most complex life cycles, a three-phase cycle. In *Polysiphonia* male and female sex organs are borne on separate plants. Observe the male sex organs, *antheridia,* densely crowded on the lower branches. The female sex organ, *archegonium,* contains the egg, which is fertilized in place. The zygote divides to form a complex structure, the *cystocarp,* which contains *carpospores.* Observe the plants bearing cystocarps. The carpospore develops into a plant that also produces spores. This second generation of spores develops into a sexually reproducing plant. Examine other examples of red algae.

4. *Diatoms.* The cell walls of diatoms are impregnated with silica. They are abundant in fresh water as well as in marine environments. Because the cell walls persist after death, vast deposits, known as diatomaceous earth, have been formed through geologic time. Diatomaceous earth is used as a fine abrasive.

B. The Fungi

The fungi are a grade of organization characterized by being nonchlorophyllous, with unicellular or filamentous bodies and heterotrophic nutrition. The plant body is called the mycelium; its individual filaments are *hyphae.* The cell walls are of chitin and may also have cellulose or other polysaccharides. Fungi may have extracellular digestion. Many fungi have spore-producing hyphae that are raised above the remainder of the mycelium. These "fruiting" bodies are frequently more spectacular than the vegetative mycelium and are the parts of fungi with which most of us are familiar; e.g., puff balls and mushrooms.

1. *The algal fungi.* This group receives its name from the resemblance of its hyphae to algal filaments. The hyphae are multinucleate. Many are aquatic and some cause diseases such as potato blight.

Rhizopus is a black mold on bread. The plant forms erect *sporangiophores* that bear *sporangia.* Each *sporangium* forms *spores.* It was in this mold that *heterothallism* (the plant body of algae and fungi is also called a thallus) was first demonstrated. That is, different plants may be of different mating types although no morphological differences are evident. When hyphae of different mating types grow in close proximity, sexual reproduction occurs.

2. *The sac fungi.* The simple yeasts, the blue-green mold *Penicillium,* the cup fungi, and the parasitic powdery mildews belong to this group. *Penicillium* is an especially important genus. Some species cause food spoilage, other species are important in the manufacture of certain cheeses, and *Penicillium chrysogenum* is the source of the drug penicillin.

Examine a section through the cup of *Peziza* or some similar cup fungus to observe the characteristic structure of this group. Following sexual reproduction, a sac or *ascus* forms which usually contains eight ascospores. Examine the other demonstrations of sac fungi.

3. *The club fungi.* The parasitic rusts and smuts, the puffballs, mushrooms, and shelf fungi belong to this group. The characteristic structure can be seen in the section through *Coprinus.* The *basidium* is a clublike hypha which bears four *basidiospores* at its apex. Meiosis occurs in the basidium so that the spores are haploid. Although a sexual process occurs in the majority of club fungi, differentiated organs of sexual reproduction are not developed. Observe the other demonstrations of club fungi.

4. *The lichens.* Lichens are symbiotic associations of algae and fungi. The algae are blue-green or green and the fungi are usually sac fungi but sometimes are club fungi. The lichens are classified according to their fungal components. Little is known concerning the physiological relationships of the alga and fungus of a lichen. Examine the demonstrations of different types of lichens.

C. Protozoa

These organisms traditionally have been considered a phylum of animals, but probably can better be thought of as a level of organization in which there is a single protoplast without cell boundaries. Because of this lack of cell boundaries, the protozoa have been called either unicellular or acellular.

Protozoa are heterotrophic; many are parasitic, causing diseases such as malaria, sleeping sickness, and amebic dysentery. There is no satisfactory classification of protozoans, but the free-living species may be broadly grouped according to their locomotory structures.

1. *Rhizopods.* The *Amoeba* is an example of this group, which moves by means of pseudopodia. Many members secrete a skeleton. Observe the demonstrations of rhizopods.

2. *Flagellates.* The flagellum is the characteristic organelle of locomotion in this group. Observe the activity of *Peranema,* especially noting the activity of the flagellum.

3. *Ciliates.* These organisms feed and move by means of cilia. The presence of two or more nuclei is characteristic. Typically the micronucleus functions in sexual reproduction and the macronucleus controls most other activities. The ciliates appear to have the highest degree of cytoplasmic specialization among the protozoa.

Examine prepared slides of *Paramecium* demonstrating *fission,* a type of asexual reproduction, and *conjugation,* a type of sexual reproduction that is essentially the same as the process in *Spirogyra*. Also observe living *Paramecium*.

Examine a drop of water from an aquarium or other source and look for different kinds of protozoans.

D. The Slime Molds

These nonphotosynthetic organisms share certain characteristics with the rhizopods and others with fungi. Thus they are frequently classified in the animal kingdom or the plant kingdom. It seems more appropriate to place them in a group by themselves. Slime molds are particularly interesting organisms for the study of cellular differentiation and organismic development. Examine the demonstration material available.

3. METAPHYTA

The Metaphyta are the green terrestrial, multicellular plants that have multicellular reproductive organs and embryonic development. All metaphytes have alternation of generations.

A. Bryophyta

This phylum consists of small plants that lack vascular tissue. In addition, these plants are dependent on water for reproduction because the sperm swim through a film of water from the antheridia to the archegonia. The conspicuous generation is the *gametophyte* or n generation, which forms gametes. The *sporophyte* or 2n generation is more or less inconspicuous and is dependent on the gametophyte and forms spores.

1. *The liverworts*. Liverworts usually grow in moist places close to the ground. Observe the gametophytes of *Marchantia* or some other liverwort. The antheridia and archegonia develop on stalked outgrowths in this genus. A given gametophyte bears either antheridial or archegonial stalks. Examine the demonstration slides of sections through antheridia and archegonia. The sporophyte is fairly simple and small. Observe the median longitudinal section through the sporophyte.

2. *The mosses*. Mosses are generally not as dependent on moist conditions as liverworts. The sporophyte is more conspicuous and more complex.

Observe the demonstrations of the life history of a moss, of various kinds of common mosses, and of sections through antheridia and archegonia. In mosses, the sexes may be separate or on the same plant.

B. Tracheophyta

This phylum is characterized by the presence of vascular tissue. The sporophyte is the conspicuous generation.

The phylum can be divided into two or more subphyla. These subphyla can be placed into two groups: a group of seedless plants and a group of seed formers.

1. *Seedless tracheophytes.* Examine the following demonstration materials:

   a. Club mosses. *Lycopodium* and *Selaginella.*
   b. Horsetails. *Equisetum.*
   c. Ferns. Fern life history and dried sporophytes.

Note that club mosses and horsetails have small or reduced leaves and that ferns have a large leaf which is the dominant organ of the plant. Ferns are more widespread than the other seedless tracheophytes. For the most part the seedless plants today are remnants of an ancient, more widespread and abundant flora.

2. *Seed plants.* The dominant vegetation of the earth's surface today is composed of plants that form seeds. These plants may be divided into two large groups: the *angiosperms* that bear seeds inside of fruits and that have flowers, and the *gymnosperms* that have seeds exposed or naked on the surface of an appendage. Although the seeds of gymnosperms may occur in cones or strobili so that the seeds are hidden from view, they are not enclosed in a fruit.

A. GYMNOSPERMS. Gymnospermous plants are either trees, shrubs, or vines. Pine, fir, hemlock, juniper, cedar, spruce, cycads, cypress, and ginkgo. Observe the demonstration of common gymnosperms.

Examine the life cycle of the pine. All seed plants form two kinds of spores in different sporangia. *The megasporangia* are surrounded by several layers of cells and are called ovules; one functional megaspore forms in the ovules. *The microsporangia* form microspores which are shed as pollen grains; the male gametophyte is contained within the pollen grain. The functional megaspore within the ovule is *never* shed but completes its development into the female gametophyte within the ovule. The embryonic sporophyte develops within the female gametophyte still within the megasporangium (ovule). When the embryo has developed, the ovule is called a seed. Thus the gametophytic generations are greatly reduced and occur within structures of the previous sporophyte and give rise to new conspicuous structures of the succeeding sporophyte.

B. ANGIOSPERMS. The flowering plants dominate the world's vegetation. They are the most important food plants for man. Cereal grains have formed the base upon which civilizations have been built.

Examine the cross section through the ovary of *Fritillaria* and observe the female gametophyte, the ovule, and the megasporangium. Correlate the structures mentioned in alternation of generations with the parts of the flower.

4. METAZOA

All Metazoa are multicellular, but within this multicellularity are various grades of organization.

A. The Sponges

The sponges appear to be a side branch of animal evolution unrelated to other metazoans. They are characterized by the presence of *pores* and *canals* that permeate the tissue. All sponges are aquatic and the adults are sessile. The supporting structures are either spicules or spongin.

Examine a simple, vase-shaped sponge. Note the large central cavity. Also observe one of these sponges that has been halved. Note the numerous pores lining the internal cavity. These pores are the inner ends of canals in the wall of the sponge.

Also examine the demonstration slides of skeletal material and the other sponges.

B. The Acoelomates

These animals do not have a true body cavity or coelom. The only cavity is the cavity of the gut which may also serve as a transport system. This level of organization is exemplified by *Planaria*. The whole mount shows how the gut branches to all parts of the body as a means of distributing food. In these small animals, gaseous exchange occurs throughout the body surface. The cross section illustrates the solid nature of the body wall.

The acoelomates are represented by two major phyla, the Cnidaria, characterized by having stinging cells, and the Platyhelminthes, the flatworms.

1. *Cnidaria.* The jellyfishes, corals, sea anemones, and hydroids (including the freshwater hydra) are commpnly called coelenterates. They are mainly marine. Observe the various examples of this phylum.

2. *Flatworms.* Two classes of flatworms are entirely parasitic. Observe a fluke and a tapeworm. Note that these organisms have greatly proliferated reproductive structures as an adaptation to their mode of life.

C. The Pseudocoelomates

Animals at this level of organization have a body cavity, but it develops differently than the true body cavity and differs in the nature of the lining of the cavity. The cross section of the roundworm *Ascaris* illustrates this organization. This is a simple gut plus reproductive organs in the cavity and a muscular body wall surrounding the cavity. The cavity probably serves as a transport mechanism.

A number of phyla occur at this level of organization. Roundworms, rotifers, and horsehair worms are some of the representatives. Examine the living rotifers. They typically have groups of cilia at their anterior ends with which they set up a current of water. Small particles of food are filtered from the current of water.

D. The Coelomates

All the remaining metazoans have a true body cavity although it is developed to varying degrees in different phyla. The cross section of the earthworm exemplifies this type of organization. Especially notice the thin layer of epidermal cells lining the body cavity; this layer is characteristic of coelomic animals.

1. *The annelids.* This phylum consists of worms in which the body is divided externally and internally into *segments.* Leeches, earthworms, tube worms, and clam worms are examples.

2. *The mollusks.* Clams, snails, chitons, squid, octopus, cuttlefish, slugs, mussels, and tooth shells are some of the representatives of this phylum. All mollusks are based on a similar body plan: a ventral muscular *foot;* a *visceral mass,* dorsal to the foot, and which contains most of the body organs; a *mantle,* a tissue layer that encloses the visceral mass and the gills and which in most mollusks secretes a calcareous *shell.* All mollusks are aquatic except for a few snails and their allies. Observe the living snails and other demonstrations.

3. *The arthropods.* Members of this phylum are found in all environments in which life occurs. One class, the insects, has more species of organisms than all the rest of the metazoa together. The arthropods are segmented and have jointed appendages. The basic organization was studied in the dissection of the crayfish. Some other members of this phylum are the mites, ticks, centipedes, millipedes, spiders, scorpions, horseshoe crabs, true crabs, lobsters, shrimp, and barnacles.

Using the crayfish as your standard, examine the various demonstrations and determine how body organization and appendages vary among the various arthropods. (See Fig. 19-1.)

4. *The echinoderms.* The starfish, sea urchins, sea cucumbers, sea lillies, and brittle stars all have a water vascular system. Externally this system is evident in the rows of tube feet. All echinoderms are marine. The name echinoderm stems from the presence of a "spiny skin." Examine a starfish, a sea urchin, and a sea cucumber. From your study, determine the characteristics shared by all three and note their differences. (See Fig. 19-2.)

5. *The chordates.* All chordates have a notochord and gill slits at some time in the life cycle. The "lower" chordates are marine and many are sessile. The gill slits are evident in the tunicate and the lancelet.

The vast majority of the chordates are vertebrates. Vertebrates are found in most environments. They and the insects are the most successful terrestrial animals. Their general body plan was examined in the dissection of the frog.

The vertebrates are divided into seven classes: the lampreys, cartilaginous fishes, bony fishes, amphibians, reptiles, birds, and mammals. Examine the available material and see if you can find structural features that are distinctive for each group.

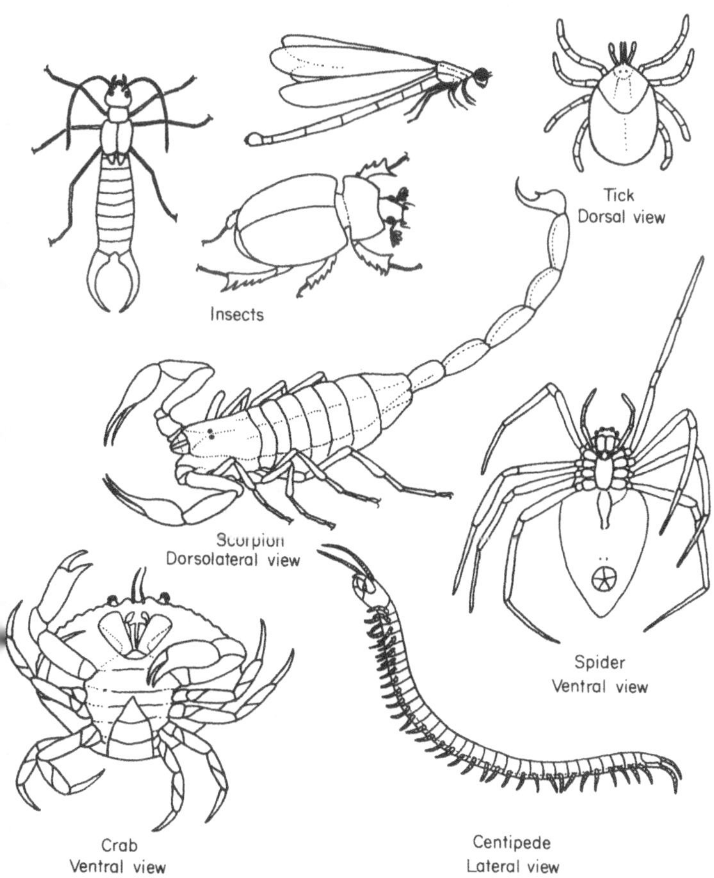

FIG. 19-1. Representative Arthropoda.

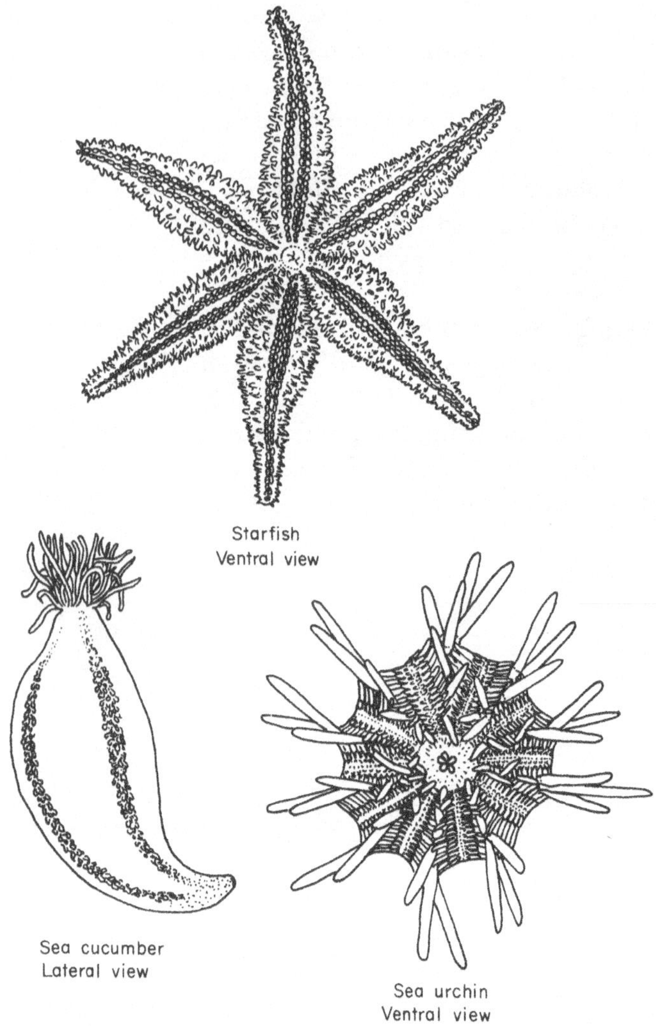

Fig. 19-2. Representative Echinodermata.

If time permits and if you so wish, you may dissect any of the animals that we have surveyed. Contact your laboratory instructor and he will provide you with the material.

SUGGESTED READINGS

Alexopoulos, C. J. and Bold, H. C. (1967). "Algae and Fungi," 135 pp. Macmillan, New York, Paperback.

Bold, H. C. (1964). "The Plant Kingdom," 118 pp. Prentice-Hall, Englewood Cliffs, New Jersey. Paperback.

Delevoryas, T. (1966). "Plant Diversification," 143 pp. Holt, New York. Paperback.

Doyle, W. T. (1965). "Nonvascular Plants: Form and Function," 147 pp. Wadsworth, Belmont, California. Paperback.

Fingerman, M. (1969). "Animal Diversity," 184 pp. Holt, New York. Paperback.

Hanson, E. D. (1964). "Animal Diversity," 118 pp. Prentice-Hall, Englewood Cliffs, New Jersey. Paperback.

APPENDIX

1. THE METRIC SYSTEM

The metric system of measurement is customarily used in scientific work to express quantities. The system is simpler to use than the English system because the metric system uses powers of ten to show the relation between different units of the same quantity. The units of the metric system are formally defined by an international body responsible for maintaining standards. Most nations have adopted the metric system, but proposals to that effect have not succeeded in the United States. Thus we must deal with the problem of converting from one system to the other. Some of the basic units of the metric system and some of the useful equivalents are presented below.

Common Fractions			Common Multiples		
fraction	prefix	symbol	multiple	prefix	symbol
10^{-1}	deci	d	10	deka	da
10^{-2}	centi	c	10^2	hecto	h
10^{-3}	milli	m	10^3	kilo	k
10^{-6}	micro	μ	10^6	mega	M
10^{-9}	nano	n	10^9	giga	G

The prefixes may be used with any of the units of measure presented below. However, a few examples are given to clarify usage.

A. Volume

The basic unit is commonly called a liter (l), but is defined as a cubic decimeter (dm^3). A liter of water at its maximal density weighs almost 1 kg. Therefore, one ml (equal to 1/1000 l) of water weighs, for most practical purposes, 1 gm. One cubic centimeter (cm^3 or cc) is 10^{-3} dm^3. Therefore, for most purposes, one cubic centimeter equals 1 ml and the units ml and cc are commonly used interchangeably. In practice, it is preferable to use the units that actually are measured.

B. Temperature

Although commonly called centigrade, the basic unit is the Celsius degree, °C. 0° C is the freezing point of water; 100° C is the boiling point. Celsius and Fahrenheit (°F) may be interconverted by using the following equation: $9/5\ °C = °F - 32$.

C. Length

Basic unit is the meter (m), which equals 39.37 inches.

Common Multiples and Fractions

kilometer (km) = 10^3 m
centimeter (cm) = 10^{-2} m
millimeter (mm) = 10^{-3} m
micron (μ) = 10^{-6} m
millimicron (mμ) or nanometer (nm) = 10^{-9} m
angstrom (Å) = 10^{-10} m

D. Mass

Basic unit is the gram, which equals 0.03527 oz.

Common Multiples and Fractions

kilogram (kg) = 10^3 gm
milligram (mg) = 10^{-3} gm
microgram (μg) = 10^{-6} gm

E. Equivalents

1 in. = 2.54 cm	1 cm = 0.3937 in.
1 lb = 453.6 gm	1 Kg = 2.20 lb (avoir)
1 US fluid oz. = 29.57 ml	1 ml = 0.033814 US fluid oz.
1 US liquid quart = 0.946 l	1 l = 1.05671 liquid quarts
1 cu. in. = 16.386 cm^3	1 cm^3 = 0.061 cu. in.
1 sq in. = 6.4516 cm^2	1 cm^2 = 0.1550 sq. in.

1 mm^3 = 0.001 cm^3 (cc) = 10^{-9} m^3 = 0.001 ml

2. DETERMINATION OF DISSOLVED OXYGEN IN WATER USING THE WINKLER METHOD

The photosynthetic rate of aquatic plants such as *Elodea,* the unicellular algal *Chlamydamonas,* or others can be determined by measuring the amount of oxygen produced by the photosynthesizing plants that is dissolved in water, and comparing that value to a control.

Experimental plants will be placed in 20-ml test tubes which are filled to the top with an appropriate medium and stoppered with the complete exclusion of air. To eliminate all air bubbles, tilt the completely filled tube to one side and insert the rubber stopper firmly. Conduct the experiment. After the experiment is completed, the following procedure is used to measure the amount of O_2 produced in each test tube:

A. Adding Winkler Reagents

Manganous sulfate and an alkali containing sodium iodide are added to the water samples. This produces a brown precipitate which absorbs all the free oxygen in the water. The sample is then acidified with sulfuric acid, which causes the release of free iodine in an amount precisely equivalent to the amount of free oxygen that was in the sample.

Procedure:

The first three Winkler reagents are added to the test tubes without removing the plants. The reagents are in glass dropper bottles; the proper volume to add is scratched on each dropper, near the point where the dropper bottom begins to narrow. Consider whether the contents of all tubes should be tested simultaneously or consecutively.

1. To each test tube add 0.2 ml of manganous sulfate.
2. Add next 0.2 ml of alkaline iodide below the surface of the water sample. Then dip this dropper in and out of a vial of distilled water so it is clean before returning it to iodide bottle.
3. Stopper each test tube with a rubber stopper introduced into the angled tube so no air bubbles remain inside. It's important that *NO* free air remains in the tube.
4. Mix contents by occasionally inverting each tube for approximately a minute. Allow the brown precipitate to settle for a few minutes.
5. Unstopper and add 0.2 ml of sulfuric acid to the side of the tube. *Use caution handling the acid.* Wipe test tube with paper towel if sample runs down the outside. Restopper and mix by inverting until the brown precipitate dissolves and the yellow free iodine is evident. (Continue on, step 6.)

B. Titrating with Sodium Thiosulfate

The purpose of this step is to measure the amount of free iodine now present. The sample is titrated with sodium thiosulfate which binds the free iodine:

$$2\ Na_2S_2O_3 + I_2 \rightarrow Na_2S_4O_6 + 2\ NaI$$

A starch indicator is also added to the sample; its blue color, indicating free iodine, disappears when all the iodine is bound as titration proceeds.

Procedure:

6. After step 5 is completed, let the algae settle. Decant from each test tube 10 ml of the water sample (measured with a 10-ml graduated cylinder) into another 20-ml test tube, with the algae or *Elodea* remaining in the first tube.
7. Add a few drops of starch solution from the glass dropper bottles to each 10-ml sample.
8. Titrate each 10-ml sample with a 0.01 N sodium thiosulfate solution, contained in the amber dropper bottles: Add the sodium thiosulfate a drop at a time. After each drop is added, mix the sample by stoppering and inverting. Count the drops carefully.
9. Continue to titrate, until the blue color JUST disappears. Once this point is reached, ignore any subsequent recoloration. Record the total number of drops used. Each thiosulfate dropper has a number marked on the rubber bulb, which indicates the number of drops equal to 1 ml of liquid. Convert each sample's number of drops used into milliliters.

C. Calculations

The oxygen content may be calculated from the following formula:

$$\text{ml } O_2/\text{ml} = \frac{N_{\text{thio}} \times 5.6 \times V_{\text{thio}}}{\text{sample size}}$$

Where N_{thio} equals the normality of the sodium thiosulfate, which in this experiment is a normality of 0.01. V_{thio} equals the volume (in milliliters) of the sodium thiosulfate titrated. Because the *sample size* is 10 ml, ml O_2/ml of sample may be determined by multiplying the values in the top line and moving the decimal point one place to the left.

If one knows the volume of the water in the test tube in which the experiment was conducted, the total amount of oxygen in the water may be calculated by multiplying the value determined from the above formula by the volume of water in the test tube.

REFERENCES

Hoar, William and Hickman, Cleveland (1967). "A Laboratory Companion for General and Comparative Physiology," pp. 286–289. Prentice-Hall, Englewood Cliffs, New Jersey.

Standard Methods for the Examination of Water and Waste Water, 12th ed., pp. 405–410. 1965, Am. Public Health Assn., New York.

Made in United States
Cleveland, OH
24 September 2026

44759665R00118